"A clinician's guide that does just that
experience of working with people with obsessive-compulsive disorder (OCD), you will find helpful ideas and pointers to guide your understanding and practice. If you're new to working with OCD, this book has all you need to get started…and if you're an experienced clinician there's a wealth of new ideas to refine your expertise. Van Niekerk draws together the most salient aspects of cognitive behavioral therapy (CBT) in its various forms to provide a concise overview of good practice in the treatment of OCD. I like the combination of theory, clinical examples, and practical advice. A great addition to the therapist's shelf!"

—**Elizabeth Forrester, PsychD, CPsychol, AFBPsS**, consultant clinical psychologist specializing in the treatment of OCD with an independent practice in London, UK; and author of *How to Deal with OCD*

"*A Clinician's Guide to Treating OCD* by Jan van Niekerk is a clear, readable, and comprehensive guide to current empirically based assessments, conceptions, and treatments for OCD. Each treatment is presented concisely and critically, with all its subtleties and distinctiveness accurately conveyed and amply illustrated through the eyes of a seasoned scholar of the field who is also a master clinician. A must-read, up-to-date guide for clinicians, academics, and students alike."

—**Kieron O'Connor, PhD, MPhil, CPsychol, AFBPsS, FCPA**, director of the OCD Spectrum Study Centre at the University Institute of Mental Health at Montreal, and full professor in the department of psychiatry at the University of Montreal

"This highly engaging text guides clinicians through five of the most empirically supported psychotherapies for OCD. Each intervention is described in a manner accessible to novice clinicians yet provides insights that seasoned therapists will learn from. A wealth of case examples, transcripts, and worksheets makes this a practical and helpful guide for treating OCD."

—**Robert Hindman, PhD**, faculty at the Beck Institute

A CLINICIAN'S GUIDE *to* TREATING OCD

The Most Effective CBT Approaches *for* Obsessive-Compulsive Disorder

JAN VAN NIEKERK, PhD

New Harbinger Publications, Inc.

Publisher's Note

Distributed in Canada by Raincoast Books

New Harbinger Publications, Inc.
5674 Shattuck Avenue
Oakland, CA 94609
www.newharbinger.com

Cover design by Amy Shoup

Acquired by Jess O'Brien

Edited by Rona Bernstein

Indexed by James Minkin

Library of Congress Cataloging-in-Publication Data on file

20 19 18

10 9 8 7 6 5 4 3 2 1 First Printing

Contents

Foreword

Obsessive-compulsive disorder (OCD) presents unique challenges to the therapist. It is highly heterogeneous (as quoted in this book, "If you have seen one person with OCD, you have seen one person with OCD," p. 1), the internal logic of the OCD system can be elusive to both client and therapist, and the most effective treatment demands of people that they engage in the very behaviors against which they have developed elaborate and highly ingrained avoidance rituals. Treatment can be a difficult sell, and many people feel unable to face it, dropping out at the start. Others attempt treatment but are unable to progress beyond a certain point. Still others simply do not seem to respond to treatment. As such, the actual effectiveness of psychological treatment for OCD remains at a stubborn and unacceptable rate of about 50 percent.

Treatment of OCD presents numerous pitfalls for therapists who have not had a lot of experience with OCD, even if they are highly experienced in other areas. The distress that individuals experience is extremely high, but at the same time, so is the perception of the necessity to engage in the very behaviors that contribute to that distress.

To therapists whose experiences in treating OCD have been unrewarding, *A Clinician's Guide to Treating OCD* will be most welcome. This book delivers a crisp, clear, well-informed, and balanced overview of OCD presentation, assessment and formulation, and cognitive-behavioral therapy approaches. Each chapter on treatment offers a highly accessible explication of the theoretical principles underlying the model and a thorough but concise description of treatment strategies.

Written in an authoritative and reassuring tone that will benefit new and experienced therapists alike, this book presents a simple—but not simplistic—understanding of OCD, why it persists, and how to help people overcome it.

—Christine Purdon, PhD, CPsych

Introduction

If this book's title caught your attention, I assume you must be interested in learning more about psychological therapy for obsessive-compulsive disorder (OCD). Perhaps you are a psychotherapist trainee or a more experienced therapist seeking to update yourself on the newest developments. The odds are that you have heard of a behavioral therapy method called exposure and response prevention (ERP) treatment or have experience in administering this. ERP, developed in the 1960s, represented a revolution in the treatment of OCD and continues to be the gold standard. When participating in ERP, clients are encouraged to expose themselves to distress-provoking items or situations while holding back from performing rituals or compulsions.

However, given that OCD is such a wide-ranging condition with such unique presentations ("if you have seen one person with OCD, you have seen one person with OCD"), it is now recognized that ERP has certain limitations: not every person benefits equally. This is why cognitive-behavioral therapy (CBT) practitioners try their best to also understand the *cognitions* (or thought content and processes) that contribute to OCD. They try to improve on ERP, offering new perspectives and modifications of therapy. The best-developed CBT approach emphasizes the contribution of the person's misinterpretation of intrusive thoughts, experienced by most people, with or without OCD, as suggesting danger or threat. (For clarity, this particular model, and the therapy based on it, will hereafter be referred to as cognitive therapy (CT), even though it falls under the broader heading of CBT.) However, modern CBT encompasses various approaches, and recent developments have seen the introduction of three newcomers: inference-based therapy (IBT), metacognitive therapy (MCT), and acceptance and commitment therapy (ACT).

Faced with this jumble of jargon and acronyms, you might understandably feel a bit bewildered. The problems presented by individuals with OCD can be perplexing, and many clinicians encounter frequent obstacles when administering the standard treatment. Although new directions within CBT add considerable value to the field, these latest developments can seem to add to the complexity.

About This Book

This book seeks to inform you about, clarify, and carefully lay out the different routes to OCD solutions. It provides an introduction to treatment approaches, old and new, as well as a comprehensive tool set to flexibly tailor treatment to each individual. The emphasis is on equipping and updating clinicians at all levels of experience with an overview of the insights that modern CBT has to offer in the treatment of OCD. The focus is on adults, although there is considerable overlap in the treatment model for adult, adolescent, and pediatric groups.

Chapter 1 provides an overview of OCD characteristics. It will update you on the DSM-5 diagnostic criteria for OCD and help you learn to distinguish OCD from related conditions and identify associated disorders. The chapter also presents a synopsis of psychological and psychiatric theories and treatments. Chapter 2 comprehensively describes the intake process in which you gather information about the client's history, followed by the psychometric assessment of OCD, which will allow you to decide on a diagnosis and make a treatment recommendation.

Chapters 3 through 7 cover the nuts and bolts of various CBT approaches: ERP, CT, IBT, MCT, and ACT. For each, I will outline the treatment model and then discuss the OCD assessment and treatment procedures that derive from the model. The aim is not to pedantically set treatment approaches apart, but to convey their main ideas. Given their theoretical overlap, you will inevitably see some degree of clinical intersection and blending.

At the end of the book, you will find several appendices with forms and worksheets that you can use in your practice. These are also available to download from the website for this book, http://www.newharbinger.com/38952. (See the very back of this book for more details.)

How to Use This Book

If you are new to CBT, I suggest reading chapters 1 and 2 to gain the necessary OCD background before proceeding to the established treatment approaches described in chapters 3 and 4. This will allow you a sufficient grounding to consider the newer approaches in chapters 5 through 7, which do not need to be read in sequence.

If you are a seasoned CBT practitioner, you can afford a more flexible approach. I hope that chapters 1 and 2 will refresh your memory and introduce you to new developments in diagnosis and assessment. You can then dip into chapters 3 through 7 according to your preference.

Whatever your level of experience, knowledge, and expertise, I hope you enjoy the book and that it will benefit and enrich your practice.

CHAPTER 1

Characteristics, Theories, and Treatments

Chapter 1 presents the key background information a practitioner needs to master in order to effectively administer OCD treatment. The following topics will be considered: (1) diagnostic and symptom features and how OCD can meaningfully be divided into subgroups; (2) demographic features, such as prevalence, course, impact, prognosis, and co-occurrence with other disorders; (3) how to distinguish conditions that overlap with OCD; and (4) neuropsychiatric and psychological theories and treatments.

Symptoms and Characteristics

The sections below will define obsessions and compulsions and how they relate to each other, provide OCD diagnostic criteria, and describe OCD subtypes.

Diagnosis

The *Diagnostic and Statistical Manual for Mental Disorders, Fifth Edition* (DSM-5; American Psychiatric Association [APA], 2013) classifies OCD as one of several overlapping disorders in the category of "Obsessive-Compulsive and Related Disorders." OCD diagnostic criteria are presented in the text box below. (For a description of other disorders in the same category, see "Conditions to Distinguish from OCD.")

As the label suggests, the central diagnostic feature of OCD is the presence of obsessions and compulsions.

Obsessions are defined as repetitive and persistent thoughts, urges, or images that the individual experiences as unwanted and intrusive and that cause significant anxiety, distress, discomfort, disgust, or other negative feelings such as embarrassment, guilt, or shame. (The Latin root of "obsess" is *obsidere*, which means to occupy or besiege; from the Online Etymology Dictionary. This testifies to the experience of many people with OCD of being "besieged" by their condition.)

Compulsions, or rituals, are repetitive behaviors or mental acts that are performed in response to an obsession and are aimed at preventing a feared consequence, reducing anxiety or discomfort, neutralizing or defusing the obsession, or achieving self-reassurance.

The compulsion is a voluntary act that follows the precipitant of an obsession. However, many people with OCD describe an evolution in their awareness of the obsession over time, particularly if the compulsion is longstanding, at which point it can become a habit that is performed almost automatically (apparently without any initial thought about why they are performing the compulsion). Nevertheless, the implicit basis for the compulsion can usually be easily identified by asking the client what he expects to happen if he faces the obsession-triggering situation without doing the ritual. He may describe an external event (e.g., "The house might be burgled") or a cognitive or emotional consequence (e.g., "I won't be able to think of anything else and will feel awful for days"). The habit-forming nature of rituals can be explained to clients by using the analogy of driving: initially every driving decision was carefully and intently thought about (clutch in…gears…gas…clutch out); however, with increasing experience, decisions are made rapidly and frequently, without conscious awareness.

It is important to carefully consider the motivation for OCD ritualizing, the *why* (or function) in addition to the *what* (content); this is a central feature of psychological models of OCD. Traditionally, in OCD, in tandem with the anxiety disorders, *harm avoidance*, or the fear of negative consequences (e.g., contaminating someone else with germs, the house being flooded, or lashing out impulsively) has been emphasized. However, another motivation is now recognized as significant: *incompleteness* or "not-just-rightness"—the emotional sense that actions, intentions, or experiences have been incompletely achieved, which manifests most directly in so-called precision or "just so" OCD, described below. In this and other forms of OCD, this may act as an alternative or supplementary motivating factor to harm avoidance (Summerfeldt, 2004).

DSM-5 Diagnostic Criteria for OCD (abbreviated; APA, 2013)

A. Presence of obsessions, compulsions, or both:

   Obsessions are recurrent and persistent thoughts, urges, or images that are experienced, at some time during the disturbance, as intrusive and unwanted, and that in most individuals cause marked anxiety or distress. The individual attempts to ignore or suppress the obsessions or neutralize them with some other thought or action (i.e., the compulsion).

   Compulsions are repetitive behaviors (e.g., handwashing, ordering, checking) or mental acts (e.g., praying, counting, repeating words silently) that the individual feels driven to perform in response to an obsession or according to rigid rules. The compulsions are aimed at preventing or reducing anxiety or distress or preventing some dreaded event or situation; however, they are not connected in a realistic way with what they are designed to neutralize or prevent, or are clearly excessive.

B. The obsessions or compulsions are time-consuming (take more than one hour per day) or cause clinically significant distress or impairment in important areas of functioning.

C. The obsessive-compulsive symptoms are not attributable to the physiological effects of a substance (e.g., a drug of abuse, or medication) or another medical condition.

D. The disturbance is not better explained by another mental disorder (see "Conditions to Distinguish from OCD" below).

Specify if:

With good or fair insight: The person recognizes that his or her OCD beliefs are definitely or probably untrue, or may or may not be true.

With poor insight: The person thinks the beliefs are probably true.

With absent insight/delusional beliefs: The person is completely convinced the beliefs are true.

Specify if:

Tic-related: The person has a current or past history of a tic disorder.

Level of Insight

There is growing recognition that some individuals with OCD have compromised insight into the reality of their fears, which resulted in changes to the DSM-IV and DSM-5 allowing the clinician to specify the level of insight (see previous text box). Low insight may manifest in different ways: a small minority of clients may have an enduring, high level of conviction that their fears are accurate and valid (described as *overvalued* ideation), although much more commonly, individuals with OCD consider their fears unwarranted when they are in the clinic setting or outside of the obsession-triggering situation but experience stronger ambivalence when their anxiety is activated in the situation itself.

Subtypes

OCD is a wide-ranging condition, and the symptom presentation closely reflects people's individual concerns and life histories. Even experienced practitioners may find themselves faced with obsessions they have never encountered before. However, despite the wide variation in individual symptoms, certain broad themes (or subtypes) have been distinguished. The most common subgrouping method considers symptom dimensions, within which the *content* of obsessions logically links to the *function* or intended aim of the compulsions or rituals.

OCD can also be subdivided according to other features, such as age of onset (e.g., childhood versus adult), comorbidity (e.g., with or without a tic disorder), or biological features (e.g., PANDAS, or pediatric autoimmune neuropsychiatric disorders, associated with streptococcal infection—a group characterized by the onset or worsening of obsessional symptoms during a childhood infection with streptococcal bacteria).

The question of which subgrouping method and which subgroups represent the most useful division (for example, in terms of having different causes and perhaps benefiting from different treatments) is the topic of ongoing research. An extensive list of symptoms is provided in the Yale-Brown Obsessive-Compulsive Scale (Y-BOCS; Goodman et al., 1989) Symptom Checklist and as part of the Dimensional Yale-Brown Obsessive-Compulsive Scale (DY-BOCS; Rosario-Campos et al., 2006; details provided in chapter 2). In my view, the broad groupings described below, starting with more common presentations, give an adequate account of OCD categories usually encountered in practice.

CONTAMINATION

Contamination obsessions represent fears about having been contaminated or the possibility of contamination with a wide range of agents, such as germs (e.g., HIV, STDs, bird flu), toxic chemicals (asbestos, lead), or disgusting substances (urine, feces, sweat). Feared contamination can also be abstract, such as being "mentally polluted" (see Rachman, 2004), morally corrupted, or otherwise changed in an undesirable way, for instance after presumed contact with or proximity to an "immoral" person (e.g., after an encounter with a married person who tried to commit infidelity, or after possible contact with a criminal or prostitute).

Common safety behaviors (strategies the person uses to try to feel safe) include ritualistic cleaning, decontamination, checking (e.g., carefully checking whether an object to be used is in fact clean), and avoidance of "risky" items, situations, or people. Individuals with contamination obsessions frequently describe a complex chain of contamination—A (dirty object or person) might have touched B, which touched C, and therefore C could be dirty—with a detailed recollection, sometimes lasting years, of which items or places may have been contaminated.

CHECKING

The checking subtype of OCD is focused on a fear of causing harm to oneself or others, such as by making mistakes—leaving the door unlocked, the oven on, the phone charger left in the socket, the taps open, or not logging out of one's computer—that potentially lead to upheaval or calamity (e.g., burglary, fire, flooding, data theft). The person, fearing she will be responsible for such an outcome, tries to allay her fears by performing multiple checks; these may form part of an elaborate and set routine (e.g., check each switch three times, say "off" aloud at the end, or feel the window latch and then look at the window from several vantage points), and, when interrupted, she has to start again at the beginning. She may also check the newspaper for reports of hit-and-run accidents ("Maybe I knocked over a cyclist without realizing it").

Individuals with a checking subtype of OCD might use other strategies to try to reassure themselves, such as taking photos with their mobile phone as additional proof that they have checked, or even retaining the item to be checked in their possession (such as taking hair straighteners to work). The "checking" label for this subtype is perhaps misleading, as ritualistic checking features are present across subtypes, but in this group it is at the fore of the presentation, also in terms of the client's complaints.

AGGRESSIVE AND SEXUAL

In this subtype, the person fears acting impulsively or aggressively (such as stabbing a vulnerable person, blurting out insults, or doing or saying something embarrassing or inappropriate, like swearing at a funeral service) or having an experience or acting in a way that is sexually inappropriate or inconsistent with the person's values or identity (e.g., being or becoming a pedophile, being sexually deviant, or being gay when straight, or straight when gay). In response, the person may ruminate (repeatedly go over the same material) in order to gain reassurance that the fears are unfounded or avoid obsession-triggering situations (e.g., avoid contact with sharp objects when in the company of others, or not hug nephews and nieces for fear of unwanted experiences).

The obsessions can apply to fears about past or future actions (e.g., "I worry that I might have done/will do something sexually inappropriate to my child, such as grope him, even though I can't remember such an act and have no intention of doing it"). Confessing past misdeeds or troubling experiences to others—even reporting these to the police—is a common reassurance-seeking strategy, but any gains tend to be short-lived. The emotions of anxiety, guilt, and shame predominate.

SOMATIC

Somatic obsessions involve the feared possibility of getting an illness, such as venereal disease or AIDS, despite a lack of objective evidence to justify concern, in response to which people attempt to reassure themselves by checking their health status or seeking reassurance from others. In the context of OCD (rather than health anxiety; see below), they do not generally experience bodily symptoms of illness but may worry that past actions put them at risk.

PRECISION, SYMMETRY, AND EXACTNESS

These obsessions represent excessive concerns about objects or actions not being perfectly symmetrical, aligned, "just so," precise, tidy, in a state of completeness, or suitably achieved (e.g., books organized exactly according to size, or letters written exactly the same distance apart). The person seeks to attain the desired state by ordering items (e.g., ensuring all the pens in the box face the same direction), repeating actions (retying shoelaces to be perfectly matched), introducing additional actions (after performing a turn, also rotating in the opposite direction, to "unwind the tension"), counting (having to count to a certain number before commencing an action), or transforming numbers (having to correct "wrong" numbers by addition or subtraction to a "right" number). There may be a rationale of magical thinking (e.g., something terrible might happen to a loved one if the item or action isn't "just right" or the right feeling state is not attained) or a motivation of perfectionism or habit ("I must get all the tasks done on my to-do list, regardless of priority; otherwise things will fall apart"). The rituals are time-consuming, frustrating, and generally unacceptable to the client (i.e., ego-dystonic).

The person may have become preoccupied with attaining strict noncommonsensical criteria as a goal in itself, resulting in slowing of her behavior. In the extreme, this may manifest in a rare condition labeled *primary obsessional slowness*, in which the person performs everyday tasks, mainly self-care activities, in an exceptionally meticulous manner and sequence (e.g., shaving or brushing teeth in an exceedingly slow, painstaking fashion). This incapacitating condition is characterized by a relative absence of anxiety (at least if rituals are not interrupted), fear, or other cognition—sometimes referred to as being cognitively "hollow" (see Rachman, 2003a).

RUMINATIONS

Rumination, in the context of OCD, reflects active, time-consuming, elaborated compulsive thinking in response to questions such as *Do I really exist?* (also called *existential obsessions*) or questions with little relevance to the person's genuine interests or ambitions, such as *How has that building been engineered?* Ruminations can be distinguished from more typical, brief, cognitive (or mental) compulsions, carried out according to preformulated requirements; ruminations tend to be more effortful, variable, and open-ended, with spontaneous thoughts introduced into the compulsive flow (see De Silva, 2003).

RELIGIOUS

Obsessions with religious themes concern the possibility of blasphemy or sacrilege, excessive concerns about morality, or not having observed a religious ritual such as prayer or confession in the proper and correct way. The response to the obsession may involve several safety strategies, including attempting to suppress the offending thoughts, neutralizing them (e.g., by reversing actions to get to a point in time prior to the offending thought), or avoiding obsession-triggering situations (e.g., attending religious institutions, walking past a church).

Despite research studies supporting the categories listed above, you may sometimes encounter symptoms that do not neatly fit any one category. In that case, to establish the diagnosis, consider the nature of the symptoms (thoughts, emotions, sensations, and actions, and what *form* they take, e.g., repetition) and the *functional relationship* between them (what is the problem, how is it dealt with, and what is the outcome). Then consider the degree of correspondence between these data and the diagnostic model of OCD. It would be worthwhile to look at the Y-BOCS Symptom Checklist (Goodman et al., 1989), described in chapter 2, to get a sense of the breadth of presentation.

Demographic Features

The sections below describe how commonly OCD occurs in the population; patterns of treatment seeking; the course, impact, and prognosis of OCD; and whether OCD tends to run in families.

Prevalence

The OCD lifetime prevalence rate (i.e., the percentage of a population that has presented with OCD at any time up to the point of assessment) is estimated at between 2 and 3% (about 1 in 40), and this tends to apply across national regions (see Eisen, Yip, Mancebo, Pinto, & Rasmussen, 2010). However, subclinical OCD symptoms are much more common—28% of a large US sample experienced obsessions or compulsions at some point in their lives (Ruscio, Stein, Chiu, & Kessler, 2010).

OCD is slightly more common in women than men. Nevertheless, males usually predominate in child and adolescent OCD cases, which may be a consequence of males' tending toward a younger age of onset. Despite a typical onset in the late teens in males and early twenties in females, onset before puberty is not rare (Eisen et al., 2010). Grant et al. (2007) found that only 11% of their sample developed OCD at or after the age of thirty.

Accessing Treatment

Many people with OCD conceal their symptoms from others out of embarrassment, shame, or fear. It is not uncommon for treatment to be sought only after years of having experienced disruptive symptoms. Unfortunately, there may be a sizeable group who never present to mental health professionals, or who are never properly diagnosed. The latter may be contributed to by health professionals who are overly focused on contamination and checking OCD (rumination subtype is particularly likely to be missed).

Course, Impact, and Prognosis

The great majority of individuals with OCD have a chronic condition, with some symptom fluctuation but without clear-cut recovery or deterioration. An episodic course, with periods of symptom flare-up separated by periods where symptoms resolve completely, occurs in between 10 and 15% of clients. About 6 to 14% have a deteriorating course (Eisen et al., 2010).

The World Health Organization (2001) estimated that in 2000, OCD was among the top twenty causes of illness-related disability for people in the age group fifteen to forty-four years. At the mild end of the spectrum, individuals might complain of only mild distress and minor impediments in getting on with their day; at the most severe end, they may be unable to attend to basic activities of daily living, such as preparing food, maintaining personal hygiene, or leaving the house. They may have developed a high dependency on the support of others, and their symptoms may encroach on family life, which can lead to rancor and contribute to relationship breakdown.

Prevalence in Families

The majority of studies have found that OCD tends to run in families: there is an increase (up to five-fold) in the occurrence of OCD among the first-degree relatives (parents, children, or siblings) of people with OCD, compared with a nonclinical control group (see Pauls, Abramovitch, Rauch, & Geller, 2014). This increase is more pronounced in the case of childhood-onset OCD. Monozygotic (identical) twins of individuals with OCD have higher rates of OCD than dizygotic (nonidentical) twins (Browne, Gair, Scharf, & Grice, 2014). These findings point to a genetic contribution to the disorder.

Comorbidity

OCD has a high level of co-occurrence with a range of other mental health symptoms and disorders, particularly anxiety disorders (e.g., social phobia, specific phobia, panic disorder) and mood disorders (e.g., major depressive disorder, bipolar spectrum disorders; Ruscio et al., 2010).

When associated with depressive symptoms, OCD can be the primary diagnosis—when it is clinically more prominent and causing greater interference than depression—with depressive symptoms *secondary* to the obsessional symptoms, reflecting the demoralization, hopelessness, and frustration that can be caused by struggling with OCD. However, depression can also be the primary disorder, with secondary obsessional features—where the onset of OCD symptoms follows the onset of depression, and OCD symptoms resolve when the depression clears up. Or, in a small minority of cases, the two conditions may simply co-occur and be relatively independent of each other.

Conditions to Distinguish from OCD

Many disorders manifest in repetitive thoughts or behaviors and may appear similar to OCD. This can cause spirited debate among clinicians about which is the correct diagnosis. Adding further scope for confusion, psychiatric terminology is continuously evolving; a label such as "compulsion" may have designated something entirely different in the past, and its use according to the old definition may have prevailed among some professionals. The sections below will seek to clarify the distinction between other disorders and OCD.

Depression

Depressed clients experience recurrent thoughts with themes of helplessness, hopelessness, and negative evaluation (e.g., *I am a failure*, *My situation is hopeless*). However, unlike obsessions, the content of thoughts changes frequently, they do not reliably elicit compulsions or avoidance, they are defended rather than resisted, and the dominant emotional tone tends to be one of pervasive depressive mood rather than anxiety or unease. Generalized behavioral limitation and social self-isolation tend to be more common depressive features.

Generalized Anxiety Disorder (GAD)

Both conditions feature anxious worry, but in GAD this concerns everyday problems and concerns, such as work, relationships, or finances, and tends to be wider-ranging than a narrower OCD obsessional focus. OCD ritualizing tends to deviate more obviously from a realistic strategy in terms of its aims, as opposed to trademark GAD—prolonged worry, aimed at foreseeing danger and preparing for mishap (e.g., "If I think all the possibilities through carefully, this will allow me to be prepared").

Health Anxiety

It has been said that people with OCD fear *getting* an illness, whereas those with health anxiety (older term: *hypochondriasis*) fear *having* an illness, despite reassurance that their fears are unfounded. There may be some basis for this distinction; however, the boundary is not clear-cut—people with contamination obsessions sometimes describe thoughts about the possibility of having contracted an illness. Both OCD and health anxiety feature overestimation of threat, leading to repetitive checking and reassurance seeking (see Reuman et al., 2017).

However, there are differences on close examination. Health anxious fears may be based on a misinterpretation of bodily sensations (e.g., where a mild body ache is interpreted as suggesting cancer). People with health anxiety are more likely to consider their fears to be realistic, and these fears are solely illness-focused, whereas people with OCD are more likely to also experience obsessions in nonhealth domains (see Hedman, Ljótsson, Andersson, Rück, & Andersson, 2017).

The distinction is complicated by new diagnostic categories in DSM-5: the previous category of hypochondriasis has been subdivided into somatic symptom disorder (fear of having an illness based on misinterpretation of bodily symptoms) and illness anxiety disorder (where there are few or no bodily symptoms and the person is preoccupied with being ill). Therefore, illness anxiety disorder appears to overlap more with OCD, although people with this condition are more exclusively preoccupied with fears about health.

Tic Disorders, including Tourette's Disorder

These conditions, under Neurodevelopmental Disorders in DSM-5, are characterized by repetitive, stereotyped (or ritualistic) behaviors, including motor or vocal tics (movements or sounds), which individuals experience as mostly involuntary and difficult or impossible to resist. Examples of simple motor and vocal tics are eye blinking, mouth movements, shoulder shrugging, throat clearing, sniffing, and tongue clicking; examples of complex tics are touching objects, twirling, jumping, echolalia (repeating others' words), palilalia (repeating one's own words), and coprolalia (use of obscene language). Tics typically have an onset in early childhood and usually improve by the late teens. They tend to be precipitated by a sensory sensation or urge, and they act to reduce tension, unlike compulsions, which are aimed at preventing feared outcomes or neutralizing concerns (see O'Connor, 2005).

Addictive Disorders and Impulse-Control Disorders

Repetitive problem behaviors are present in conditions such as pathological buying or gambling (listed in DSM-5 as gambling disorder), kleptomania, and pyromania. However, these are aimed at attaining gratification, exhilaration, or tension-reduction, whereas OCD compulsions are aimed at preventing feared outcomes and relieving distress or discomfort.

Disorders Related to OCD

In the DSM-5, OCD falls under "Obsessive-Compulsive and Related Disorders." These related disorders include body dysmorphic disorder, hoarding disorder, trichotillomania, and excoriation disorder.

In body dysmorphic disorder, there is a preoccupation with an imagined defect in the person's appearance (e.g., in features of face, skin, hair, build, genitalia) or an excessive and disruptive apprehension about an existing anomaly. The person responds to this concern with compulsive or avoidant behaviors, such as mirror checking (or avoiding) or excessive grooming (e.g., applying excessive base make-up to hide a defect in the texture of skin). A person may even seek cosmetic surgery to remedy the perceived defect. An additional or alternative diagnosis of OCD is considered when the focus of obsessions is not restricted to appearance.

Hoarding obsessions and compulsions previously were considered a subtype of OCD, but on the basis of observed differences with OCD, DSM-5 classifies hoarding disorder separately. The main features are excessive acquisition of and failure to discard items with little or no apparent value, resulting in living areas becoming cluttered and nonfunctional; distress is centered mainly on letting go of items. Compared with OCD, the person is less likely to consider the behaviors to be a problem (i.e., the disorder tends to be *ego-syntonic*) and the treatment prognosis is poorer. Additionally, there are neurocognitive differences between hoarding disorder and OCD (Mataix-Cols et al., 2010). Although OCD can also manifest in having clutter, this is for a different purpose than hoarding, such as retaining potentially contaminated items so as not to risk exposing others to harm.

Interestingly, even though some studies have found poorer performance on measures of various neuropsychological abilities in people with hoarding disorder compared to healthy controls, a recent study comparing *unmedicated* clients with hoarding disorder with controls found no differences other than in the types of strategies used in a perceptual categorization task (Sumner, Noack, Filoteo, Maddox, & Saxena, 2016). This points to a common problem with neuropsychological studies in patient groups, including OCD, posed by possible confounding effects of medication use.

Trichotillomania and excoriation disorder are respectively characterized by excessive hair-pulling (common areas include the scalp and eyebrows) or skin-picking (e.g., picking at healthy skin, scabs, or bumps, even to the point of bleeding or injury). Hair-pulling can follow an *automatic* pattern, where the behavior occurs in the absence of awareness while the person is engaged in another activity, like reading or watching television. Or it can manifest in a *focused* pattern, where the activity is intentional and the center of attention (see Flessner et al., 2008). However, these behaviors are not precipitated by an obsession causing distress, as in OCD; therefore, they serve a function of tension relief or gratification, rather than anxiety reduction or preventing feared outcomes. Both conditions are more common in women, whereas OCD is more equally distributed across the sexes.

Obsessive-Compulsive Personality Disorder (OCPD)

Confusion here stems from overlapping names. OCPD reflects a longstanding personality style of excessive devotion to work, perfectionism, rigidity, moralistic attitudes, and overcontrol; however, these are viewed as consistent with the person's values (ego-syntonic) and not considered problematic in their own right. In contrast, OCD symptoms are experienced as distressing and problematic, resulting in treatment seeking.

Theories, Models, and Treatments

Over recent decades there has been substantial progress in psychiatric and psychological theories and treatments of OCD, making different claims about the factors that determine the vulnerability to and maintenance of OCD. The question of how these different sets of hypothetical causal factors relate to each other is considered next, followed by a separate discussion of each.

So What Causes OCD?

This is a common question from clients. The exact set of causes of OCD has not been definitively established, but contributing roles for genetic, neurobiological, environmental, behavioral, and cognitive factors have been suggested. One can distinguish the factors that predispose an individual to OCD from the factors that precipitate (or activate) and maintain the condition. The preconditions that have to apply for the disorder to take its onset define the *vulnerability*. In OCD, genetic, biological, and psychological factors (e.g., childhood adversity, stressful life events, past learning) likely confer the vulnerability. Precipitating factors may include the stresses associated with major life events such as pregnancy. Both neurobiological and psychological factors, such as unhelpful thinking processes and behaviors, contribute to maintaining the condition; these can be addressed through medication, psychological therapy, or both. Some studies (see Linden, 2006) have found overlapping changes in the brain as a consequence of CBT or medication for OCD, so there may be convergence in the treatment mechanisms despite different methods being used. The following section will consider a detailed view of neuropsychiatric theories and treatments.

Neuropsychiatric Perspectives

A neurochemical theory of OCD was put forward on the basis of findings that OCD responds preferentially to antidepressant medications (so called because they are an established treatment for depression) that act to modify the effects of serotonin, a neurotransmitter, or chemical messenger, used by certain neural *circuits*, or interconnected nerve cells (see Zohar, Greenberg, & Denys, 2012). This category of antidepressant is referred to as selective serotonin reuptake inhibitors (or SSRIs). The view was that OCD was caused by abnormalities in the neural circuits that used serotonin as a neurotransmitter. However, it is now thought that there may be abnormalities also in other neurotransmitter systems, such as dopamine and glutamate (Pauls et al., 2014). Dopamine plays a role in reward-motivated behavior and the control of movement, and glutamate is the major *excitatory* neurotransmitter; in other words, its release increases the likelihood that a nerve signal will be transmitted.

Researchers of serotonin in OCD face complexity when interpreting their findings because many people with OCD have depressive symptoms, and depression can fuel OCD; therefore, they also have to consider the explanation that SSRIs may not benefit OCD directly but act through reduction of depressive symptoms. A further difficult question is posed by the fact that not all people with OCD benefit from SSRI treatment.

Other methods used to investigate brain processes in OCD include structural neuroimaging (to investigate brain structure, e.g., MRI scanning), functional neuroimaging (to examine physiological processes in the brain, e.g., fMRI or PET scanning), and cognitive testing (administering pencil-and-paper or computerized tests of mental abilities such as memory, attention, and IQ).

Neuroimaging studies in OCD have revealed abnormal brain activity in regions such as the orbitofrontal cortex (at the front of the brain—important for high-level, complex executive thinking processes including inhibition of behavior and emotional control), the caudate nucleus (a structure of the basal ganglia in the center of the brain—important in goal-directed action), and the anterior cingulate cortex (involved in emotional regulation, error monitoring, and other abilities). This has led to the *frontostriatal* model being formulated (see Pauls et al., 2014), which proposes that an imbalance between feedback loops leads to hyperactivity of the pathway connecting the orbitofrontal region and subcortical areas of the brain. Consequently, exaggerated, anxious concerns result in persistent conscious attention to the perceived threat (culminating in obsessions) followed by compulsions aimed at neutralizing the threat; the compulsions are reinforced by bringing relief from anxiety.

This is an oversimplification, but it can be helpful to explain the frontostriatal model to clients in the following way: "When danger is sensed, the brain switches on a warning light. Usually the warning light would be switched off after it has been established that there is in fact no danger; however, in OCD, the process for switching off the warning light does not work as effectively as it should, leaving the person with a lingering sense of danger and a feeling that he needs to do more to make the situation safe, despite there being no realistic need for that."

The field of *epigenetics* concerns processes that activate or deactivate genes. Pauls et al. (2014) put forward a neuroepigenetic model and speculated that OCD triggers (such as adverse events around childbirth, psychosocial stressors, trauma, and inflammatory processes) influenced the activity of genes related to neurotransmitter systems (such as serotonin, dopamine, and glutamate), contributing to dysfunction in frontostriatal brain regions and making the person vulnerable to OCD.

Neuropsychological Perspectives

Neuropsychological studies have tried to establish whether there are underlying neuropsychological dysfunctions that account for the symptoms of OCD. There may be particular difficulties in abilities such as memory (therefore explaining doubting obsessions and checking) and the ability to suppress thoughts or actions, inappropriate to the context, also called *response inhibition* (thereby accounting for the repetitive nature of obsessions and compulsions). Some studies (reviewed in Menzies et al., 2008) have tried to establish whether abilities supported by frontostriatal areas of the brain were selectively affected in OCD (see previous section for a description of the frontostriatal model).

In a thorough review of all the studies comparing the performance of people with OCD with healthy control participants, Abramovitch, Abramowitz, and Mittelman (2013) found evidence of poorer performance in participants with OCD on various neuropsychological tests. The effect sizes (a statistic that reflects the size of the difference) for various abilities are shown in the table below.

Effect Size	Cognitive Ability
Medium to large	Memory, particularly memory for nonverbal material (e.g., designs) more than for verbal material (e.g., word lists)
Medium	Executive or higher-level abilities (including planning ability, cognitive flexibility, and response inhibition) Sustained attention Speed of thinking
Small	Working memory (the ability to hold and manipulate information in mind, such as repeating a list of digits backward, or repeating a sequence of spatial locations tapped on a board, backward) Visuospatial abilities

The authors speculated that the poorer nonverbal memory performance was a likely consequence of poorer organizational ability (which falls under the category of executive function) rather than a memory problem in itself. Overall, the impairments observed in OCD clients were moderate; *they all fell short of attaining the threshold that in neuropsychological practice would be considered a meaningful impediment to functioning* (Abramovitch et al., 2013).

The authors cautioned that researchers had not considered that different OCD subgroups might show differences in neuropsychological ability and that possible effects of medications had not been carefully investigated. Also, as neuropsychological impairments have been observed in many psychological disorders, too few studies have compared OCD with other client groups.

Finally, given the cross-sectional nature (providing a snapshot of functioning at one point in time) of the vast majority of neuroimaging and neuropsychological studies, it is not easily possible to conclude whether the observed differences with healthy controls are a cause or a consequence of OCD. Future painstaking longitudinal research (preferably following a group of people over time, some who develop OCD) will be necessary to provide a clearer picture of the neurobiological, psychological, and environmental factors that contribute to the causal pathways in OCD.

Treatment with Medication

It is useful for psychological practitioners to have a working knowledge of psychiatric medications, as these are frequently co-administered with psychological treatments. Often, the psychological therapist has the most regular contact with the client. It is a routine part of my assessment to ask clients how they are doing with their medication. However, if the client has questions or concerns about the medication or the dosage, these should be directed to the responsible medical practitioner.

The first-line treatment in adults with OCD, which has been shown to be effective, is the SSRI class of antidepressant medications, including citalopram, escitalopram, fluoxetine, sertraline, paroxetine, and fluvoxamine. The alternative is clomipramine, which is a tricyclic antidepressant—an earlier class than the SSRIs. A common characteristic of these medications is that they increase the effects in the brain of the neurotransmitter serotonin (Fineberg & Craig, 2010; Grant, Chamberlain, & Odlaug, 2014).

The general recommendation is to prescribe the medication and incrementally increase the dosage to the higher end of the allowed dose range (although many benefit from a lower dosage; see Bloch, McGuire, Landeros-Weisenberger, Leckman, & Pittenger, 2010). It has been thought that the anti-obsessional effect takes longer to develop than an antidepressant effect and therefore treatment should preferably be continued for up to twelve weeks to establish whether there is a benefit. However, this view has been drawn in doubt by a recent review of SSRI studies in adults, finding that a significant benefit tends to occur already after two weeks and that, on average, 75 to 80% of the treatment gains (compared to a placebo intervention) are evident within six weeks (Issari, Jakubovski, Bartley, Pittenger, & Bloch, 2016).

If one SSRI does not work, the psychiatrist may prescribe another or recommend a switch to clomipramine. If there still is no benefit, an antipsychotic medication (so called because their first use was in the treatment of psychotic disorders; they are also called *neuroleptic* medications) can be added, such as risperidone or aripiprazole (see meta-analysis by Dold, Aigner, Lanzenberger & Kasper, 2015). These may help by reducing the effects of dopamine in relevant areas of the brain, although aripiprazole has a more complex action on dopamine (see Padder, 2015). The exact mechanisms are unresolved.

Recently, there has been a lot of interest in the possibility of boosting the effects of exposure and response prevention therapy for OCD (chapter 3) by administering a medication called D-cycloserine (DCS) shortly before or after exposure to obsession triggers. (DCS is an antibiotic drug that was developed to fight tuberculosis but has been found to stimulate a brain receptor thought to be involved in the consolidation of new memories.) It was hoped this would encourage anxiety reduction over the course of exposure and diminish recurrence after. A recent meta-analysis of treatment trials (where results from all the different studies are pooled) investigating the use of exposure therapy for post-trauma symptoms, anxiety, and OCD did find a small benefit—less than 4% symptom improvement—for participants who received DCS compared to those who received a placebo (Mataix-Cols et al., 2017). This benefit diminished when the two groups were compared at follow-up, after treatment had ended.

The advantages of opting to take medication for OCD are that it is evidence-based and convenient, but this is balanced by disadvantages, including the possibility of side effects. Common SSRI side effects include nausea, headache, insomnia, and sexual difficulties, such as delayed orgasm or lack of libido, and these tend to be exacerbated at higher dosages. Initial medication side effects may improve over time as the person's body adjusts to the medication, or a different medication in the same or a different class can be tried.

The side effects of clomipramine are usually more marked than for the SSRIs, which is why the latter are usually tried first. Common side effects of clomipramine are sleepiness, dizziness, constipation, urinary retention, dry mouth, blurred vision, and weight gain. Another disadvantage is that the drug is considered more dangerous in overdose than the SSRIs.

The side-effect profile of the antipsychotic class of medication is less favorable than for the antidepressants and can vary considerably between different medications. (More information can be obtained from sources such as Padder, 2015. Also see the comments on the possibility of obsessive-compulsive symptoms associated with the use of the drugs clozapine and olanzapine in the context of schizophrenia treatment, in "History of OCD" in chapter 2.)

Another disadvantage of pharmacotherapy for OCD is a high risk of symptoms worsening, or the person having a full relapse, when treatment is stopped. This problem can be reduced by taking medications long term, but that may incur additional risks, which require management by a medical professional experienced in OCD, preferably a psychiatrist. Sometimes a low drug dosage is sufficient to maintain improvement, with the benefit of minimizing side effects and reducing the risk of possible untoward consequences of long-term treatment.

A further issue is the presence of residual (ongoing) symptoms despite improvement. To put this into perspective, SSRIs produce *only* a 20 to 40% improvement in symptoms (although this is still impressive) over a twelve-week period for most people with OCD (Abramowitz & Jacoby, 2015).

How do medication and CBT (see next section) compare? Hirschtritt, Bloch, and Mathews (2017), in their review paper, provide the standardized mean difference (a statistic that here reflects the extent of symptom reduction over treatment, compared to that in the comparison group): 0.61 for SSRI medications, 0.95 for clomipramine, and 1.33 for CBT/ERP (see psychological treatments discussed next). (A common rule of thumb is that 0.5 and 0.8 constitute a "medium" and "large" effect, respectively.)

Psychological Perspectives

Psychological theories of OCD have a long history (see Jenike, Baer, & Minichiello, 1998). In some of the earliest accounts, people with obsessions, particularly blasphemous or sexual obsessions, were considered to be possessed by the devil, and the treatment of choice was exorcism, designed to drive out malign spirits.

William Shakespeare wrote the play *Macbeth* in 1606, depicting Lady Macbeth's guilt-ridden handwashing after she was complicit in the killing of the king. Did she have OCD? In my view, the quality of Shakespeare's psychological insights, as revealed in his writing, classes him as a first-rate psychologist!

The religious explanation eventually shifted to a medical view. The first description of obsessions and compulsions in the psychiatric literature was in the writings of the French psychiatrist Esquirol in 1838. By the end of the nineteenth century, obsessions were considered an expression of depression.

In the twentieth century, the view shifted toward a more psychological explanation. The French psychologist Pierre Janet described the treatment of compulsions with what we now refer to as behavioral techniques. With the advent of Freudian psychoanalysis, the contribution of unconscious conflicts was put forward (in 1909 Freud published a famous case study he called "Rat Man," which described his analysis of a client with obsessions of harm befalling loved ones); however, this approach was limited by its lack of scientific credentials and convincing evidence that treatment was effective. Today, psychoanalytic therapy is not considered evidence-based treatment for OCD.

BEHAVIORAL MODELS AND ERP TREATMENT

With the development of behavior therapy in the 1950s, the insights that were helpful in the treatment of phobias were also applied to OCD. The behavioral theory of OCD was based on Mowrer's two-stage theory of fear and avoidance (see Mowrer, 1960; Rachman & Hodgson, 1980). In the first stage, obsessional fears develop through a process of *classical conditioning*, where objects or thoughts acquire the ability to elicit anxiety because of their association with an aversive experience. Hereafter the person's response of performing compulsions or avoidance of obsession triggers is maintained through *negative reinforcement* (i.e., reduction of an unpleasant experience—in this case, anxiety).

The behavioral model led to the development of a game-changing treatment for OCD called exposure and response prevention (ERP), first described in a 1966 case study by the British psychologist Victor Meyer. In brief, ERP involves two steps: first, the client is exposed to triggers of obsessional fears (e.g., touching "dirty" doorknobs), and then the client is helped to refrain from any avoidance or escape response (no washing afterward). Repeated exposures aim to reduce anxiety or discomfort through a process of *habituation* (or "numbing out"), although newer developments in behavioral theory question the importance of habituation (see discussion of inhibitory learning theory in chapter 3).

In the subsequent decades, extensive evidence has accumulated testifying to the indisputable effectiveness of ERP for OCD. This is currently the gold-standard psychological treatment approach for OCD and is a flagship example of evidence-based clinical psychology, of which CBT is one of the strongest models. In a meta-analysis of thirteen ERP trials, Foa and Kozak (1996) found that an average of 83% of participants were "responders" at the end of the study (usually defined as at least 30% improvement in symptoms from the start of the trial). Across sixteen studies that reported follow-up data, 76% of participants were responders at follow-up (average follow-up period was twenty-nine months; Foa & Kozak, 1996). This shows that considerable numbers of people who have ERP experience substantial benefit (although these findings reflect the outcome of therapy under the ideal conditions that tend to characterize research settings). A comparison of effect sizes for ERP versus medication is provided in "Treatment with Medication" above.

However, authors including Clark (2004) have pointed to limitations in the behavioral theory of OCD. For example, it fails to explain why many obsessions develop without an apparent pairing with an aversive experience (i.e., classical conditioning). It does not explain why obsessions tend to evolve over time and why many people have several obsessions. It fails to distinguish adequately between OCD and anxiety disorders and fails to account for certain observations, such as a reduction in obsessions in obsessional checkers when the experimenter is present in the obsession-triggering situation. The most important limitation is perhaps that ERP is not a perfect treatment: many clients, despite improvement, have continuing lower-level symptoms, and it is estimated that 25% refuse the treatment (Schruers, Koning, Luermans, Haack, & Griez, 2005) and 15% drop out of treatment (Ong, Clyde, Bluett, Levin, & Twohig, 2016). The emotion of disgust, often elicited by contamination obsessions, presents a specific challenge as exposure has been shown to be less effective for reducing disgust than reducing fear; cognitive reappraisal methods (see below) may be more useful here (see Olatunji, Berg, Cox, & Billingsley, 2017).

COGNITIVE-BEHAVIORAL THERAPIES

CBT practitioners tried to address some of the limitations of behavior therapy for OCD. They did this by addressing the "cognitive" factors thought to play a role in OCD. (*Cognitive* is an umbrella term that refers to thought content and processes, such as beliefs, rules, assumptions, reasoning, attention, thinking styles, and biases; these can be contrasted with "behavior," or observable action.) CBT is a structured therapy that aims to address the contribution of dysfunctional thinking and behavior in psychopathology.

Confusingly, there is considerable variation in the extent to which cognitive factors in OCD are emphasized in treatment protocols. ERP aimed at achieving habituation can be introduced first (as the primary component), followed by cognitive work introduced in later sessions (e.g., Steketee, 1999); other protocols adopt a more comprehensive cognitive focus, with behavioral work aimed primarily at achieving belief change rather than habituation (e.g., Wilhelm & Steketee, 2006). All of these can collectively be labeled as "CBT"; however, in chapter 4, I will describe a cognitively focused therapy approach for OCD that requires limited use of behavioral exposure methods.

CT derived from the appraisal-based model. The best-established cognitive therapy approach is that within the appraisal-based model (ABM), which considers the development of OCD to result from two key factors: (1) the misinterpretation of normal, commonly experienced intrusive thoughts (e.g., *I've had a sexual thought about the Virgin Mary, which must mean I'm evil*), and (2) the subsequent misguided attempts to control or neutralize the resulting obsession, such as avoidance, thought suppression, and ritualizing (e.g., avoiding going to church, trying to block "evil" thoughts, and saying a prayer every time an obsession is experienced to cancel it out).

A key research finding (referenced in chapter 4) has been that of "nonclinical obsessions and rituals"—that the large majority of people without OCD report intrusive thoughts and ritualistic behaviors. These intrusive thoughts differ only in quantity and severity from those of people with OCD, but not in their content.

Cognitive therapy aims to use cognitive and behavioral strategies to test two competing theories: (1) that the occurrence of certain thoughts represents danger, or (2) that it is feared that the thoughts may represent danger, resulting in the use of counterproductive coping strategies (Salkovskis, 1999). The aim is for the client to formulate an alternative, helpful, realistic interpretation of her intrusive thought and for her underlying core beliefs to be corrected, therefore removing the need to perform unhelpful safety responses.

Credible evidence has accumulated that cognitive therapy is an effective treatment for OCD, but the evidence to date has not demonstrated any advantage over ERP (see Rosa-Alcazar, Sánchez-Meca, Gómez-Conesa, & Marín-Martínez, 2008). A further challenge to the appraisal-based model is presented by findings that a notable percentage of people with OCD do not score high on measures of beliefs thought to be key issues in OCD (e.g., Polman, O'Connor, & Huisman, 2011).

Newcomers: inference-based therapy, metacognitive therapy, and acceptance and commitment therapy. Recently there have been exciting newcomers on the CBT stage, including inference-based therapy (IBT), metacognitive therapy (MCT), and acceptance and commitment therapy (ACT). (I provide detailed descriptions in the chapters to follow.)

These therapies have been developed either for OCD specifically (IBT), as a transdiagnostic protocol (ACT), or for generalized anxiety initially and with various subsequent disorder-specific applications (MCT). However, all three have been tested in OCD. There was evidence from case series (testing an intervention in the absence of a control group) and randomized controlled trials (the most rigorous test for whether an intervention works) for efficacy of IBT (reviewed in Aardema, O'Connor, Delorme, & Audet, 2017; Julien, O'Connor, & Aardema, 2016), ACT (Twohig et al., 2010; reviewed in Bluett, Homan, Morrison, Levin, & Twohig, 2014), and MCT (reviewed in Fisher, 2009; Normann, van Emmerik, & Morina, 2014; van der Heiden, Van Rossen, Dekker, Damstra, & Deen, 2016).

So, what are they about? Here is a brief primer on each:

- **Inference-based therapy (IBT):** IBT conceptualizes the clinical starting point in OCD as being an inference of *doubt* (i.e., doubting what you already know; for example, the object may be contaminated with germs, or the door may be unlocked). The origins of the doubt are in the faculty of the imagination and at the expense of the immediate reality of the situation; the person becomes absorbed in a remote possibility: what *could be* there, from the imagination, supported by a narrative, which trumps what *is* there, from perception.

  In therapy, the clinician elicits the narrative reasoning supporting the situational doubts and reflects upon it with the client. The client's various thinking devices are identified. These may involve (1) making unsupported links between different categories of information, such as confusion between two separate events ("A year ago I read in the newspaper about a woman who swallowed glass in her cereal, so there could be glass in my cereal"), and (2) inverse inference, where the conclusion precedes the observation ("Chemicals are used everywhere, so there could be some on my hand"). The therapist guides the client to compare the OCD narratives with his simple, reality-based thinking evident in other situations and develop an alternative, reality-based "commonsensical" view, which he eventually acts on.

- **Metacognitive therapy (MCT):** As the name suggests, the concept of *metacognition* is central to MCT. In short, this concerns "thinking about thinking." The metacognitive model considers that in OCD, intrusive thoughts activate *metacognitive knowledge* (beliefs about thoughts or thinking processes), and this metacognitive knowledge in turn guides maladaptive processing. For example, a person engages in extended worry after she interprets a (normal) unwanted intrusive thought of committing a murder as evidence that she might in fact have committed a murder; in this case, the underlying metacognitive belief is *Having an unwanted image of a murder means I could be the murderer.*

MCT for OCD does not target the appraisals (or interpretations) of thoughts by testing them against reality (as in ABM-CT), but rather selectively targets metacognitive beliefs about intrusions (as in the example above) and rituals (e.g., *I need to worry to get rid of these thoughts*). MCT strategies are aimed at helping clients update their metacognitive knowledge and modify their metacognitive control strategies so they view their thinking from a more benign premise.

- **Acceptance and commitment therapy (ACT):** The main treatment goals in ACT are for the client to treat thoughts as nonliteral and to not avoid them, along with the aim of increasing positive-value-guided action. ACT targets a set of central processes that contribute to OCD, including the battle against unwanted thoughts and emotions and low engagement in meaningful, value-consistent life activities. A wide range of techniques, including mindfulness exercises, is used to change the function of unwanted thoughts and emotions so that they no longer present a barrier to effective action.

This chapter aimed to give you an overview of OCD: its features, causes, and treatments. You have seen how neuropsychiatric and psychological theories underpin different treatment methods. The next chapter will provide you with guidelines for conducting the initial assessment to set you up for administering CBT.

CHAPTER 2

Initial Assessment

This chapter will consider guidelines for conducting the initial assessment of a client presenting with OCD, which usually takes between one and a half to three hours, depending on the level of complexity. Subsequent chapters will consider additional assessment strategies according to the requirements of the different treatment approaches outlined.

Goals and General Guidelines

The main goals of the initial assessment interviews are to establish an accurate diagnosis and develop a problem list, both of which inform the treatment recommendation, and to start gathering information to inform cognitive-behavioral case conceptualization. The feedback to clients should allow them to make an informed choice about treatment and encourage realistic optimism and the motivation to participate in treatment.

Therapists should be adequately knowledgeable and confident about their clinical practice to inspire trust. Clients commonly feel anxious, embarrassed, guilty, or ashamed about their symptoms and may have concealed them from others for some time. Therefore, the standard guidelines for establishing rapport in clinical interviewing apply ten times over: embody an attitude of genuineness, transparency, and collaboration; communicate an accurate understanding of the client's experience; and practice unconditional positive regard for the client as a person (as a trusted mentor once told me, never forget that the client is a *person*).

Explain the need for a detailed initial assessment and gathering of background information, and tell clients to expect assessment to continue alongside treatment. Allow sufficient time for the assessment and try not to rush the client, which may be counterproductive in the long run. Present sufficient structure to assuage the client's anxiety, and use therapy time efficiently. Even in the assessment phase, clients should come to expect that appointments will follow a predictable structure that includes:

1. providing a brief update on mental states, symptoms, and events since the previous appointment;
2. collaboratively setting and covering the main agenda points for the session;
3. agreeing on any home assignments to be completed before the next appointment; and
4. summarizing and having an opportunity for feedback and questions.

Discuss your professional guidelines regarding the confidentiality of clinical information and any concerns the client might have about this. Have an explicit policy about digital data retention and security, which are important present-day concerns.

When appropriate, reduce clients' self-stigma and shame by reassuring them that OCD is a recognized and prevalent mental health disorder and that blaming themselves for having it is neither justified nor helpful ("Criticizing yourself adds a second problem, rather than just having one—your OCD"). I find the metaphor of a conman helpful: "OCD sets a very clever trap. Many very clever, resourceful people get themselves entangled in the scam. That's why we'll have to look carefully at how it works so that we can intervene in a helpful way."

Structure

In my private practice, I conduct the assessment according to the following schedule (when time is restricted, items 1 through 3 are the priority, but assessment of items 4 through 8 can focus on the most important details and do not require the full breadth of assessment described in the sections below):

1. *Presenting problems*: assessment of the client's concerns and symptoms and their history, including OCD symptoms (OCD assessment questions are considered in more detail in the sections below)
2. *Mental state examination (MSE)*: brief items on domains including mood states, anxiety, suicidal thinking and plans, thought content (e.g., obsessions, phobias, delusions) and process, behavior (e.g., compulsions, tics), energy level, appetite, sleeping, perceptual experiences (e.g., hallucinations), cognitive function (e.g., memory and concentration), and libido over the prior week or two
3. *Personal psychiatric history:* details of treatment episodes, diagnoses, and treatments administered and their impact

4. *Family psychiatric history:* details of mental health diagnoses and treatments in biological relatives
5. *Medical history:* details of current medical conditions, current medications, or other treatment
6. *Developmental history:* relevant attainments, experiences, important relationships, and problems encountered in different life phases
7. *Premorbid functioning:* personality traits, social network, hobbies, personal strengths, and values
8. *Lifestyle:* exercise, diet, smoking, alcohol intake, addictions, and illicit drug use

Assessing OCD Symptoms and Their History

The following OCD screening questions, adapted from those found on the website of the Anxiety and Depression Association of America (https://www.adaa.org/screening-obsessive-compulsive-disorder-ocd), can be useful; any yes response requires more in-depth assessment:

1. Do you have bothersome unwanted thoughts that seem silly, nasty, or horrible?
2. Do you worry excessively about dirt, germs, or chemicals?
3. Are you frequently worried that something bad will happen because you forgot to do something important?
4. Are you afraid you will act in an unacceptable way when you really don't want to?
5. Are there things you feel you must do excessively or thoughts you must think repeatedly to feel comfortable or ease anxiety?
6. Do you clean yourself or things around you excessively?
7. Do you have to check or repeat actions many times to be sure they are done properly?
8. Do these problems interfere in your life?

Once it has been established that OCD is the likely diagnosis, more in-depth preliminary assessment can proceed according to the schedule outlined next.

Current OCD Symptoms

Conducting an assessment of the following categories will elicit detailed information about the client's OCD symptoms and their level of impact.

1. *Obsessions.* What kinds of persistent thoughts, images, or urges do you have that make you feel anxious, uncomfortable, or distressed? What are their triggers? (There can be *external* triggers, such as being given change in a shop, which triggers a contamination obsession, or *internal* triggers, such as having a memory of a sexual encounter, which triggers an obsession of being a rapist.)

2. *Behavioral and mental compulsions and other safety strategies, such as thought suppression.* For each obsession or situation presented in response to the questions in item 1 above, ask:

   How exactly do you respond to or deal with the problems posed by the thoughts/what do you do to reduce your anxiety or reassure yourself/what do you do to prevent bad things from happening? Do these responses present a satisfactory solution, or do you have to repeat them? (It may also be helpful to inquire about *rescue factors* present in a situation, which, even if the person did not actively use them, may confer reassurance. An example of this would be having an antiseptic spray ready to use if needed.)

   For mental rituals specifically: Is there anything you do *in your mind* to set things right/reassure yourself/prevent bad things from happening/undo a bad thought or action/reduce your anxiety? (Examples of mental reassurance strategies used by people with OCD are saying a prayer, thinking a lucky thought, counting, undoing one thought with another, mentally reviewing a situation, and making a mental list.)

   Do you perform any other ritualized, repetitive behaviors or mental actions that we have not discussed? What triggers them/what job do they do for you? (This may allow identification of obsessions and compulsions that have not previously been identified.)

3. *Avoidance.* Are there situations, thoughts, or activities you avoid because of your obsessional fears?

4. *Negative consequences and insight.* What do you fear would happen if you exposed yourself to feared or avoided situations without performing rituals? (It is helpful to ask this with specific reference to key obsession-eliciting situations, described by the client.) How much do you believe this from 0 to 100, (a) *now*, or (b) if you were *in the situation*? (Clients are often more invested in their fears in the obsession-triggering situation.)

It can be useful to probe whether the client believes that her fears are in fact accounted for by OCD. Is there a discrepancy between her "gut feeling" and what her "rational mind" is saying, or between what she thinks and what another person might think? If there is an absence of a non-OCD viewpoint or the client strongly defends the OCD position, this may suggest compromised insight and the possibility of overvalued ideation. (Also see the Brown Assessment of Beliefs Scale below.)

5. *OCD impact on relationships and quality of life.* Has OCD caused any problems in your relationships? Have others expressed concern about you or stepped in to help you cope with your OCD? In what specific ways do they support you? (Be particularly vigilant for mention of others taking over the client's responsibilities, participating in rituals, or providing reassurance.) Are any of your relationships in danger of breakdown because of OCD? (If so, intervention in the relationship may be a priority.) In what ways is OCD holding you back in your life? What valued activities have you given up?

6. *The client's theory of maintenance of the OCD.* How does your OCD work and what keeps it going? (This information may have relevance for the treatment recommendation; for example, if the client is strongly invested in neurobiological causation of OCD to the exclusion of psychological factors, this may point to medication being a more suitable strategy.)

History of OCD

After obtaining a thorough understanding of the client's OCD symptoms, gather information about the client's history of OCD.

1. *Onset.* When (or at what age) did your OCD symptoms begin? When (at what age) did you recognize them as being a problem? What do you remember of that time? Did any significant event or circumstance occur around the time the symptoms started?

2. *Course.* Have the symptoms changed much over time? Has their severity fluctuated? What are the factors that help things, or make them worse?

3. *Precipitant to help-seeking.* Why are you seeking treatment *now*? (This may have a bearing on the client's level of motivation. For example, if he is only presenting for treatment due to family pressure but sees little point in treatment himself, you will have to intervene to enhance motivation, or perhaps agree to delay treatment until the client's motivation has increased.)

4. *Client's understanding of the causes of his or her OCD.* What caused the OCD to develop and gain a foothold?

5. *Impact.* In what areas of your life have you previously been held back because of OCD? Are there any ways in which OCD helps you? This question may seem counterintuitive; however, it is important as part of a functional analysis to acknowledge any perceived benefit (or reinforcement) of symptoms. This also allows an assessment of insight, such as when a client acknowledges disruption but overall believes that "it (the sum of the symptoms) keeps me safe." This likely has a bearing on the client's motivation to participate in treatment.

6. *Treatments.* What previous psychological treatments have you had? How many sessions? What system of psychotherapy was used? (If the client says "CBT," inquire about the content of sessions and whether the treatment involved any exposure to fear-inducing situations, as some of what is labeled CBT isn't CBT, or may fall short of the recommended approach.) What was the outcome of treatment? It may be helpful for the client to consent to your calling previous therapists to obtain information. Record cursory details of current medications: drug names, duration of treatment, dosage, outcome of treatment, side effects, and date of last review. (It may also be useful to record brief details of past medications, although this is more relevant to psychiatric practitioners.)

   When considering the client's medication regimen, it is worth taking note of a recent review of studies in which second-generation (newer) antipsychotic medications were administered for schizophrenia (Fonseka, Richter, & Müller, 2014). The authors' findings suggest that the medications, clozapine and to a lesser extent olanzapine, might induce or exacerbate obsessive-compulsive (OC) symptoms in a minority of patients receiving treatment for schizophrenia. However, remember that there is research evidence of benefits of SSRI supplementation with several antipsychotic medications in the treatment of OCD. Nevertheless, if any client develops OC symptoms after starting treatment with olanzapine or clozapine, or if existing symptoms exacerbate, consider a possible contribution of medication to these symptoms.

Obstacles to Assessment Presented by OCD Symptoms

Clients are sometimes embarrassed or ashamed to disclose symptoms or fear that there will be serious repercussions; these concerns are particularly common in regard to sexual or aggressive obsessions. Two strategies can be helpful in dealing with this: First, you can give examples of sexual and aggressive obsessions and ask the client if he has experienced this ("Some people with OCD worry that they may act impulsively in an aggressive way, or that they may have inappropriate sexual desires or act sexually improperly. Have you experienced any of these?"). Second, you can ask the client to complete a checklist of obsessions, such as the Y-BOCS Symptom Checklist (described below), which can be helpful in two respects: (1) this can be completed in the privacy of the client's home and therefore not require direct disclosure to the therapist, which may exacerbate shame, and (2) the list of clearly labeled "OCD obsessions" can serve as reassuring psychoeducation.

Another problem that may occur is that the assessment process elicits OCD symptoms. For example, being asked to complete a questionnaire may trigger concerns about mistakes, anxiety, indecisiveness, checking, or repeating. Clark (2004) suggests strategies for dealing with this:

- Validate the client's anxiety, explaining that many people with OCD experience anxiety when answering questions or completing questionnaires during assessment. This problem is a consequence of having OCD.
- Carefully explain why all the parts of the assessment are important.
- Discuss the client's fears and work out a strategy that may lessen anxiety, such as the client reading through a questionnaire with the therapist before completing it to make sure items are understood.
- Where symptoms concern a fear of making mistakes, indecision, and checking, point out that engagement with anxiety-inducing aspects of the assessment involves facing fears and that over time this strategy can be expected to be helpful. When anxiety is very high, it may be helpful (at this early point) to provide reassurance (a) that assessment will continue throughout treatment, allowing refinement of the problem conceptualization (therefore correcting a view that therapy will be irrevocably sabotaged if one "wrong" answer is given at the earliest point), or (b) that you will carefully consider all the client's responses and point out when an answer is unexpected and needs clarification (this involves an initial transfer of responsibility to the therapist, which can reduce anxiety, but it's important to recognize that inflated responsibility and need for reassurance are part of the problem and will have to be addressed later in therapy; see chapter 4).
- Keep the assessment as brief as possible and choose measures that provide information directly relevant to case conceptualization and treatment.

Be flexible in allowing clients more time to complete questionnaires, and accept that the boundary between the background assessment and treatment may be less clearly defined than what you are accustomed to. Some initial assessment may have to be sacrificed or left until later.

- Approach problems experienced during the assessment as an opportunity for data gathering. For example, if a client takes a very long time or is resistant to completing a questionnaire, ask exploratory questions: What is holding you back? Which specific parts do you expect to be difficult and why? What do you worry might go wrong if you complete it in the normal way? Have you experienced this before? What has helped before? Allow assessment to blend into treatment by gently encouraging clients to adopt a mindset of being willing to test their anxious predictions (e.g., "I'll become paralyzed with fear") and experiment with functional coping behaviors (e.g., by setting a deadline for when to have completed the questionnaire, or not going over answers again), but be careful not to be premature and overambitious, exceeding the client's coping ability and risking treatment disengagement.

Assessing Related Factors

In addition to assessment directly related to OCD, there are many pertinent factors that must be considered regarding the client's history and functioning.

Assessing Comorbid Conditions

OCD frequently copresents with other conditions, particularly depression and anxiety disorders. The referring health professional may have already suggested additional diagnoses, or clients may describe them when asked about their presenting problems. Alternatively, these may become evident from observation during interviews (a demeanor of dejected, blunted affect with slow responses to questions would raise the possibility of depression) or from the mental state examination. It is beyond the scope of this book to provide detailed guidelines for assessment of other DSM-5 disorders. However, clearly a detailed assessment of all presenting problems is warranted to establish the applicable diagnosis (e.g., clinical depression), the level of severity and associated impediments, and the relationship to the client's OCD (primary or secondary to, or independent of, the OCD). This would form the basis for deciding on an appropriate treatment plan.

Assessing and Managing Suicidality

As discussed in chapter 1 and above, OCD is frequently accompanied by depressive symptoms. And given the prevalence of suicidal thoughts and behavior in people with depression, it is important to keep in mind suicidality when conducting an assessment. In fact, suicidal thoughts and attempts are common in OCD. A study of 293 treatment-seeking adults with OCD (overall severity was in the moderate range) found that 52% reported a history of suicidal thinking and 15% had made at least one suicide attempt (Pinto, Mancebo, Eisen, Pagano, & Rasmussen, 2006). High depression severity and hopelessness are commonly associated with an increased suicide risk.

Therefore, when working with clients with OCD, it is important to assess whether depressive symptoms are present and establish whether there is suicidal risk. Standard guidelines for managing suicidal risk apply and include having the client agree on a contract to refrain from suicide attempts, agreeing on a crisis management plan, negotiating the restriction of suicide means, increasing the level of social support, referring to a medical practitioner or psychiatrist to prescribe medication, and instilling a sense of hope (see Ellis & Newman, 1996). In extreme cases, it may be justified to breach confidentiality and initiate crisis support.

Assessing and Managing Risk in the Context of Sexual or Aggressive Obsessions

It can occasionally be challenging to distinguish OCD obsessions from thoughts realistically predicting a risk of sexual offences, self-harm, or harm to others. (Examples of overlapping content include sexually desiring minors, harming the newborn baby by putting her in the microwave, or committing suicide.) The following factors support the possibility of OCD (see Veale, Krebs, Heyman, & Salkovskis, 2009; Warwick & Salkovskis, 1990); *base your interpretation on the profile of characteristics, rather than on sole criteria:*

- Ego-dystonicity of the thoughts (the thoughts are offensive, unwanted, or inconsistent with the person's values and the person lacks a genuine desire and intention to act on the thoughts)
- Not acting on or masturbating to the thoughts (sexual obsessions)
- An absence of past behavior consistent with the thought (e.g., a history of suicide attempts, or sexual or violent offenses)

- Avoidance of obsession triggers (e.g., looking away from children, or not handling knives)
- Efforts to suppress or neutralize the thoughts
- Very frequent thought occurrence
- Marked anxiety, fear, guilt, or shame about the thoughts
- Overdisclosure to the therapist of past sexual experiences, perhaps hoping to benefit from reassurance based on the therapist "knowing everything" (in the context of sexual obsessions)
- Seeking help and mental health service referrals because the client considers his thoughts to be objectionable and fears acting on them
- Other OCD symptoms, or a history of OCD

Although not acting on or masturbating to sexual obsessions is one of the indicators listed above, clients with OCD may experience intrusive thoughts while having sex or masturbating to an adult fantasy. People with sexual obsessions frequently monitor their level of sexual arousal in specific situations to test their fears (e.g., monitor their genital area for signs of arousal when walking past a child), but attention to the genitals paradoxically may increase the level of arousal, or lead to overreporting of arousal. Further, anxious sensations may be misinterpreted as sexual arousal, and anxiety linked to obsessions has the capacity to lead to physical signs of sexual arousal in men. Therefore, self-reported sexual arousal should be considered an unreliable indicator of sexual interest.

Reassure clients and their families that intrusive thoughts with violent or sexual content are common in the general population and that the person with OCD may even be at lower risk of harming others than any other member of the public. However, if you feel uncertain, lack experience, or have a reasonable basis for concern that there may be a risk of harm to self or others, seek consultation with colleagues immediately and consider appropriate steps for referral and management.

Assessing Personal Psychiatric History

The goals are to contextualize the client's OCD symptoms and history in the larger psychiatric history. This allows an opportunity to explore the interrelationship between mental health problems in the past and whether this merits consideration in the present: "Previously your OCD led you to feel depressed; how is your mood this time?"

Assessing Family Psychiatric History

Sometimes OCD is suspected, but there is diagnostic uncertainty. Given the evidence of a hereditable component of OCD, the presence of OCD in biological relatives will be a data point in support of an OCD diagnosis.

Assessing Medical History

In the context of psychological work conducted in a medical setting, it is worth taking note that a number of serious medical and neurological conditions present with repetitive behavioral features that overlap with OCD (Menchón, 2012). Therefore, management is best considered in close liaison with medical care providers, integrated within the neurological care plan.

Assessing Developmental History

The developmental history permits an understanding of who the client is and where she is coming from—the trajectory of her development in the context of her life experience. This lets you consider the historical events and circumstances that may have relevance to the client's OCD. It allows you to make predictions about the mechanisms maintaining her current problems—*the primary focus in CBT*—but also provides a historical narrative, which likely benefits therapeutic rapport by allowing the case conceptualization to be more affirming of the complexity and richness of people's lives.

So, which history is likely to be important and worth highlighting in the assessment? Currently there is little definitive research about the historical contributions to OCD. An ideal study would follow large groups of people longitudinally, assessing them at multiple time points and comparing those who develop OCD with those who don't, to identify the characteristics that distinguish them. Ideally such a comparison should be guided by a preexisting theory. However, the current knowledge largely relies on retrospective data (i.e., asking clients about their past), and therefore there is scope for inaccuracies in what people remember and report. It is also worth keeping in mind that causal interpretation in the context of developmental psychopathology research—saying a childhood event definitely contributed to the later development of a mental disorder—is generally very complex given the typical unfeasibility of obtaining rigorous experimental evidence of a causal effect (using the experimental method, one variable is manipulated to examine the impact on another variable, under controlled conditions). Historical causation frequently involves complex causal chains where predispositions, exposures, experiences, and responses interact to facilitate the development of symptoms and eventual culmination into a full-blown disorder.

Taking into account these limitations, Barcaccia, Tenore, and Mancini (2015) speculated about which childhood experiences might contribute to create "historical vulnerability" to OCD on the basis of their review of research.

First, *nonspecific* vulnerability factors are presented by any severe adverse experience in childhood, such as traumatization, emotional neglect, and abuse. That is, these experiences increase the likelihood of having a wide range of mental disorders in adulthood, without the specific nature of the event being associated with a specific future disorder.

Second, CBT research has explored links between *specific* childhood circumstances and key beliefs or issues (see chapter 4) thought to underlie OCD: overestimation of threat, inflated responsibility (an exaggerated sense of being able to cause or prevent personally significant negative outcomes), perfectionism, intolerance of uncertainty, importance of thoughts (believing that the mere presence of a thought is important), and importance of thought control (overvaluing the importance of complete control over intrusive thoughts and believing this is possible). Although this research references the appraisal-based CBT model (chapter 4), it likely has relevance to the other CBT approaches as well.

Perfectionistic beliefs are thought to result from parental overcontrol, criticism, and high standards and from the child learning from the example set by perfectionistic parents. Careau, O'Connor, Turgeon, and Freeston (2012) found an association between perfectionistic beliefs in adults with OCD and what has been termed the experience of *sociotropy* in childhood: the child sacrificing his objectives and desires to adopt those of another person, at the expense of developing his own identity.

Further, the childhood experience of *perception of threat* (where parents' behavior contributed to instilling fear of certain behaviors or situations) was associated with the OCD belief of overestimation of threat. A feeling of responsibility for the happiness or protection of one's parents or others was linked with the OCD belief of inflated responsibility. The experiences of perception of threat and sociotropy played a part in most of the assessed obsessional beliefs, suggesting that they were more general vulnerability factors.

Salkovskis, Shafran, Rachman, and Freeston (1999) hypothesized five pathways to an inflated sense of responsibility:

- Parents who encourage a broad and early-developed sense of responsibility in their children (e.g., expecting them to take on roles too advanced for their age)
- Rigid and extreme codes of behavior and duty (e.g., children internalize strict rules about appropriate behavior)

- Being shielded from responsibility (e.g., anxious parents fearfully discouraging the child from trying new experiences or taking on responsibility)
- An incident where the child's actions or inaction realistically contributed to misfortune
- An incident where it appeared to the child that her actions, inaction, or thoughts contributed to mishap (e.g., the child is angry at her teacher and wishes the teacher would disappear; the teacher is then taken seriously ill and has to leave her job)

Pace, Thwaites, and Freeston (2011) examined the role of external criticism in OCD. They attempted to draw together research findings into the following formulation: A child who experiences regular distressing parental criticism responds to this by attempting to prevent and avoid this form of punishment. This active avoidance, which entails becoming responsible for preventing situations that may elicit criticism, sensitizes the child to the notion of responsibility. The child might also adopt perfectionistic standards to try to win the approval of parents. Obsessive-compulsive behaviors may result, such as checking, for preventing danger and avoiding criticism. These may in turn attract criticism from others, leading to stress, which may perpetuate further instances of the behavior or feed into and strengthen an already inflated sense of responsibility.

In summary, when assessing the client's developmental history, ask specific questions to explore childhood experiences that may hold relevance to OCD or general psychopathology:

- Traumatization, emotional neglect, abuse
- Parental perfectionism, overcontrol, high standards
- Lack of freedom to develop one's own identity
- Being given excessive or inappropriate responsibilities
- Parental anxiety manifesting in overprotection and being shielded from responsibility
- Rigid parental codes of duty and responsibility
- Having made (or having the perception of having made) serious mistakes leading to disaster
- High parental criticism and dysfunctional methods of discipline

Questioning should explore the perception of the event(s), the personal significance and impact, and any relevance to present symptoms (e.g., How did you experience this? What did this mean to you/about you? How did you cope? How did this affect your life/behavior? What was the worst part? Do you think this links to your OCD? If yes, how does that work?) to gather the most relevant information.

Assessing Premorbid Personality

The aim here is to get a picture of the client's activities and personality functioning prior to the onset of OCD (although, if the client had OCD for a very long time, he may have a poorly defined sense of his life before his symptoms emerged). This can contribute to detailed goal setting where the overarching therapy goal is to recapture the ground that has been lost to OCD (e.g., making contact with friends he has lost touch with, reengaging with hobbies, resuming social roles), or, more ambitiously, to have a better and fuller life than before!

Values is a concept that has recently attracted renewed interest, particularly in the context of ACT (chapter 7), which stresses the importance to one's mental health of purposeful living, where behavior is closely aligned with values, though having insight into one's values has the potential to bolster interventions within any of the CBT approaches.

So what is a value? I present positive values to clients as being those lifelong characteristics embodied in their behavior (*verbally established motivation*) that would allow them to live their lives in the way that is most meaningful to them. Values are different from goals in that they are not attained by a deadline; rather, they are lifelong signposts for behavior. Some examples are kindness, honesty, integrity, persistence, self-care, resilience, and social connection. (An extensive ACT analysis of values is offered by Plumb, Stewart, Dah, & Lundgren, 2009.)

You can assess values at the start of or throughout therapy by asking clients which value-guided behaviors have been thwarted by OCD; these can inform goal setting. Further, it may be helpful to establish a framework of value "scaffolding" at the assessment stage to support and maintain motivation and reduce the risk of treatment disengagement (strategies for addressing motivational problems and working with values are further considered in chapters 3 and 7). What do clients want to stand for in the face of the challenge of treatment? Which qualities do they want to see reflected in their behavior as part of their preparation for and participation in treatment that can be tough and demanding?

Assessing Lifestyle and Health Behaviors

Assessing behaviors related to the client's lifestyle and health promotes the client's adoption of a holistic, inclusive perspective on health. Fostering appreciation of the mind-body link and pursuit of health-supporting behaviors is likely to be helpful as an auxiliary strategy in OCD treatment.

In an assessment, I always ask whether the client considers her diet to be healthy. I also ask these brief questions: Do you have concerns about your weight? Do you eat at least five portions of fruit and vegetables a day? Is your sugar intake excessive? Do you stay sufficiently hydrated?

I then ask about smoking, alcohol intake, addictions, and use of illicit drugs. Finally, I ask about her daily activity levels and whether she does structured exercise.

This brief assessment may point to the need for further diagnostic assessment, for instance of substance use disorders. It will also allow identification of an unhealthy lifestyle or health behavior deficits and present an opportunity to discuss these with the client, clarify her priorities, and establish whether she wishes to include these as (smaller) therapy goals, or whether a referral to another professional would make sense (e.g., dietitian, general practitioner, personal fitness instructor). Issues that are likely to hinder therapy progress, such as problem drinking, may have to be addressed before therapy for OCD can proceed. Other goals may be pursued alongside OCD work, such as weight reduction and increased exercising (but be careful not to be overambitious and lose focus).

Providing information can be valuable prior to considering goals. I routinely provide reading materials about diet as a home assignment (e.g., Stetka, 2016) or direct clients to websites that provide reputable information about exercise or problem drinking. I also describe the findings of research suggesting aerobic exercise may be a helpful supplemental strategy to CBT for OCD. (Abrantes et al. [2009] administered a twelve-week aerobic exercise intervention with fifteen participants with OCD and reported reductions in OCD symptoms, negative mood, and anxiety after exercise sessions; nevertheless, because this study did not include a control group, it cannot definitely be said that the exercise intervention accounted for the observed benefits.) Finally, in the spirit of data-driven intervention, it is worth taking note of the plethora of gadgets available that allow detailed behavioral assessment and recording; these may be a useful motivator for technologically minded clients!

Psychometric Assessment

The initial assessment can be complemented by the use of psychometric instruments that quantify various aspects of the client's OCD symptoms, such as symptom severity, insight, and treatment readiness (see below). These can inform the diagnostic assessment and the treatment recommendation and allow tracking of treatment progress. Psychometric assessment is also useful for depressive or anxious symptoms, although I will not discuss this further here. Instruments with application to specific CBT approaches will be considered in the relevant chapters.

Symptom Severity

A plethora of self-report instruments has been developed for assessing OCD severity. I routinely use the Y-BOCS (Goodman et al., 1989) and Obsessive Compulsive Inventory-Revised (OCI-R; Foa et al., 2002), which I have found to be quick, practical, and well tested in research, and which in my view suffice for routine clinical use. I also describe alternative instruments below, which you may consider.

YALE-BROWN OBSESSIVE-COMPULSIVE SCALE

The Yale-Brown Obsessive-Compulsive Scale (Y-BOCS; Goodman et al., 1989) has been widely used and is a gold-standard assessment in clinical trials of OCD treatments. It is a clinician-administered, semistructured interview that consists of a Symptom Checklist and a Severity Scale. A self-report version of the Y-BOCS was also developed by Baer (1991). The Symptom Checklist assesses current and past experience of more than sixty obsessions and compulsions arranged into fifteen categories:

- **Obsessions:** aggressive, contamination, sexual, hoarding/saving, religious, symmetry/exactness, somatic, miscellaneous
- **Compulsions:** cleaning, checking, repeating, counting, ordering/arranging, hoarding/collecting, miscellaneous

The Severity Scale assesses the severity of obsessions and compulsions, separately, experienced over the prior week, using five items for each (with items requiring a severity rating from 0 to 4). Each item assesses *time* occupied by obsessions or compulsions, *interference* due to obsessions or compulsions, *distress* related to obsessions or not performing compulsions, *effort to resist* obsessions or compulsions, and *degree of control* over obsessions or compulsions. The item scores are added to provide severity scores for obsessions (score range: 0–20), compulsions (0–20), and a total score that is the sum of both (0–40). The clinical breakdown for total scores, according to OCD severity, is: 0–7 (subclinical), 8–15 (mild), 16–23 (moderate), 24–31 (severe), and 32–40 (extreme).

A reduction of at least 30% in the total score indicates meaningful clinical improvement (Tolin, Abramowitz, & Diefenbach, 2005). Abramowitz (2006) pointed out limitations of the instrument: symptoms are assessed according to form rather than function, so it's up to the clinician to establish which obsessions elicit which compulsions. Also, the checklist contains only one item assessing mental compulsions. In my view, the "aggressive" obsessions category misleadingly also includes the common obsession of a fear of making a mistake and being responsible for a calamity, such as a fire.

The Dimensional Yale-Brown Obsessive-Compulsive Scale (DY-BOCS; Rosario-Campos et al., 2006) has recently been published as a successor to the Y-BOCS. The updated instrument is more comprehensive and takes considerably longer to administer. There are more data available for the older scale, which presents an advantage for interpreting scores. Therefore, the more economical Y-BOCS will likely continue to be a preferred assessment tool in clinical settings for some time to come. However, it may still be worth taking a look at the DY-BOCS, which can be downloaded from http://www.nature.com/mp/journal/v11/n5/extref/4001798x2.doc.

OBSESSIVE-COMPULSIVE INVENTORY-REVISED

I have found the Obsessive-Compulsive Inventory-Revised (OCI-R; Foa et al., 2002) scale very useful given its brevity and demonstrated sensitivity to treatment effects. It contains eighteen items, divided into six subscales of three items each according to symptom presentation: washing, checking, ordering, obsessing, hoarding, and neutralizing. Items are assessed on a scale from 0 ("Not at all") to 4 ("Extremely") and summed to produce subscale scores (range: 0–12) and a total score (range: 0–72). Abramowitz and Deacon (2006) reported that a cutoff score of 14 represented a good balance between accurately detecting people with OCD and excluding those with anxiety disorders (in their study, the probability of having OCD was 74% if the total score was 14 or higher).

DIMENSIONAL OBSESSIVE-COMPULSIVE SCALE

The Dimensional Obsessive-Compulsive Scale (DOCS; Abramowitz et al., 2010) is a practical twenty-item self-report inventory containing four subscales of five items each: (1) contamination, (2) responsibility for harm, (3) unacceptable thoughts, and (4) symmetry/completeness/exactness. This represents—in my view—a rational breakdown of OCD symptoms. Each subscale assesses symptom severity over the prior month on a 0 to 4 scale according to time occupied by symptoms, avoidance behavior, associated distress, functional interference, and difficulty disregarding obsessions and refraining from compulsions. An advantage of the scale is that every subscale is introduced with several examples of specific obsessions and compulsions in that category. The range of total scores is 0 to 80. Total scores of at least 18, and, with a higher degree of certainty, 21, suggest the presence of OCD (Abramowitz & Jacoby, 2015). The DOCS can be downloaded from Jonathan Abramowitz's website (http://www.jabramowitz.com/teaching-and-training.html).

Insight and Treatment Readiness

Clients' insight into the reasonableness of their obsessional fears and their motivation to engage in treatment are important variables to consider in making treatment recommendations and in predicting treatment success. I do not routinely administer the measures described below; instead, I informally assess the domains covered by the item content, which can serve as useful guidance for phrasing of assessment questions and areas that need to be covered. However, administering the instruments allows a more rigorous approach when this is called for.

BROWN ASSESSMENT OF BELIEFS SCALE

The Brown Assessment of Beliefs Scale (BABS; Eisen et al., 1998) is a clinician-administered, semistructured scale to rate the degree of conviction clients have concerning their beliefs. It has application to a broad range of mental disorders. First, the clinician asks the client to specify the belief, idea, or "worry" that he has been most concerned about during the prior week. Second, the client rates six items from 0 to 4 (least to most severe). (A seventh additional item that is less relevant to OCD is not included in the total score and has therefore been omitted here.) The items assessed, along with examples of interviewer questions, are:

1. Level of conviction ("How convinced are you of these ideas?")
2. Perception of others' views of beliefs ("What do you think other people think of your beliefs?")
3. Explanation of differing views ("How do you explain the difference between what you think and what others think about the accuracy of your beliefs?")
4. Fixity of ideas ("If I were to question the accuracy of your beliefs, what would your reaction be?")
5. Attempt to disprove ideas ("Over the past week, how often have you tried to convince yourself that your beliefs are wrong?")
6. Insight ("What do you think has caused you to have these beliefs? Do they have a psychological cause, or are they actually true?")

TREATMENT AMBIVALENCE QUESTIONNAIRE

The Treatment Ambivalence Questionnaire (in Rowa et al., 2014) was developed to provide an assessment of ambivalence and concern about having psychological treatment for anxiety; it was not designed specifically for OCD but finds a good application here. It consists of twenty-six items assessing the level of agreement with the possibility of various drawbacks of engaging with therapy (e.g., that treatment could be dangerous, might not be effective, and could be inconvenient and time-consuming.) This can usefully supplement the therapist's assessment of treatment readiness and motivation (also see chapter 3).

Psychometric Assessment on an iPad

A collection of self-report inventories has been compiled by NovoPsych Psychometrics (http://www.novopsych.com) for administration on an iPad. I use this application and have found it practical and useful. The client completes tests on the iPad, and the data (anonymized according to preference) are submitted automatically to NovoPsych for scoring. Within minutes, the clinician receives a report via email containing the results and interpretation.

With relevance to adult OCD, the Y-BOCS Symptom Checklist and Severity Scale (Goodman et al., 1989; see above), the Vancouver Obsessional Compulsive Inventory (Thordarson et al., 2004), the Thought Action Fusion Scale-Revised (Amir, Freshman, Ramsey, Neary, & Brigidi, 2001; see chapter 4), and the Metacognitions Questionnaire-30 (Wells & Cartwright-Hatton, 2004; see chapter 6) are available as part of the battery. It also includes measures of depressive mood and anxiety, which can be useful supplementary assessments.

Family Assessment

OCD can have a negative impact on relationships and family functioning in several ways: (1) family members may be expected to comply with rituals and avoidant behaviors (e.g., collaborate in excessive checking or cleaning, only leave the home after the client has left, or not eat in restaurants because food hygiene cannot be guaranteed), (2) family members may be asked to provide repeated reassurance in response to the client's fears, and (3) the client's negative mood states may present an obstacle to participating fully in family life or being able to attend to the needs of a partner or family member, or may instigate or exacerbate conflict.

Given the functional relevance of family interactions to OCD and well-being in general, with the client's consent, try to involve the client's partner or other relevant adult members of the household at the initial assessment stage and hear about their experience of the client's OCD symptoms. Where there are practical obstacles to their attending an appointment, a brief phone conversation may suffice. Ask questions about each of the three aforementioned issues, aiming to establish how the OCD presentation may interfere in family life, how the family members respond, and their level of accommodation or resistance.

Briefly update the client and relevant family members on the importance of having to address in treatment the ways in which family life may have reorganized itself around the OCD, thus potentially allowing symptoms to be maintained. Encourage the family not to punitively criticize or blame the client for her symptoms (family communication characterized by hostility, criticism, and emotional overinvolvement has been found to be associated with poorer treatment outcome; see Steketee & Van Noppen, 2003), and consider ways to carefully address unhelpful family responses in treatment.

Behavioral Exercises and Diary-Keeping for Data-Gathering

In addition to retrospective self-report, it can be helpful to attempt to activate clients' obsessions in the session and assess their behavior and experience. For example, in the case of contamination obsessions, depending on whether this is relevant, you can suggest the person use the clinic restroom and ask him about any obsessions and safety strategies, you can suggest modification in his safety behavior to establish whether it is within his reach and how he experiences it, or you can assess his response to your behavior (e.g., drop your pen on the floor and, without cleaning it, ask the client if he would be able to use it). I have a sink in my office, so I can ask clients to demonstrate how they wash or check the taps. Nevertheless, it is important at this early point to exercise caution, avoid springing surprises on clients, and not exceed their coping ability or willingness to tolerate discomfort.

It may also be helpful to ask clients to keep a diary of their symptoms, involving detailed sampling of at least one day of the week and one day of the weekend. A Symptom Diary worksheet is provided in appendix 1.

The Treatment Recommendation

At the end of the initial assessment, collate all the necessary information in order to make a diagnosis and consider treatment options. The two evidence-based OCD treatments are medication (SSRI or clomipramine, which can be augmented with other medications) and CBT. Their relative advantages and disadvantages are listed in the table below.

	Advantages	Disadvantages
Medication	Generally safe and simple to use 20–40% symptom reduction in most people	Residual symptoms are common The possibility of side effects A high risk of relapse when treatment is discontinued A significant minority experiences little or no improvement
CBT	Structured, time-limited therapy with average treatment duration of 15–20 sessions 60–70% symptom reduction on average (Abramowitz & Jacoby, 2015) Lower risk of relapse than medication when discontinued	Requires effort and a significant time commitment; can be uncomfortable and anxiety-inducing Complex intervention—skilled, well-trained clinicians may be in short supply

Combination treatment with CBT and medication is a legitimate strategy, but it has not been shown to be more effective than either on its own. It also poses a challenge to practitioner and client alike to developing a clear understanding of the relative efficacy of components in the treatment package in accounting for symptom benefit. According to the guidelines developed by the influential National Institute for Health and Care Excellence (NICE; 2005) in the United Kingdom, CBT is the treatment of choice for mild OCD in adults, either medication or CBT can be administered for moderate OCD, according to client preference, and combination treatment is reserved for severe OCD.

The following factors are considerations in treatment selection (see Abramowitz & Jacoby, 2015):

- The client must be able to grasp a CBT model; therefore, people with developmental disabilities or cognitive impairment may benefit more from medication.
- Clients' preferences are likely to affect treatment adherence. Therefore, if the client expresses a strong preference for one or the other after

reviewing OCD theories and the pros and cons of medication versus CBT, this preference should be strongly considered when determining the treatment approach.

- CBT is effective across the spectrum of OCD severity, but a more severe condition may require more intensive therapy (e.g., CBT combined with medication, more frequent CBT sessions, increased length of therapy).
- Clients with poor insight are considered to have a poorer treatment prognosis; therefore, CBT may require more of an emphasis on cognitive rather than exposure techniques for these clients.
- Comorbid conditions such as severe depression, bipolar disorder, generalized anxiety disorder (GAD), substance addiction, or severe personality disorders are likely to be impediments to CBT for OCD. These may have to be dealt with before commencing OCD work, using either medication, a psychological intervention, or both. In the case of depression, GAD, and other anxiety disorders, the same medications can be prescribed as for OCD, thus ideally benefiting both. Depressive symptoms *secondary* to OCD that are not sufficiently severe to contraindicate CBT are likely to improve when the client's OCD improves through treatment. With other comorbid disorders, when discussing with clients which problem they would like to tackle first, it may be reasonable to consider which problem disrupts their functioning the most.
- Clients who have not had a successful response to medication and have not had CBT are eligible for CBT.

Before giving the treatment recommendation, summarize the relevant symptom information to help the client understand the diagnosis of OCD. Provide brief, accessible information about OCD and allow an opportunity for questions and discussion of any reservations the client may have about the diagnosis; you can point the client to credible internet sources of information, such as the International OCD Foundation (which lists local and global affiliates and partner organizations). Stress that the condition is unlikely to improve in a long-lasting way without treatment and that proven treatment strategies exist, including medication and CBT. Review the pros and cons of each and give your recommendation. If the recommendation is for CBT, explain that CBT treats the factors that maintain the condition; it is not necessary for the client to have a precise understanding of the causes to be able to make progress ("to get out of a hole, you don't need to know how you got into it").

CBT goals are to more effectively manage fear-inducing situations, develop a healthier way of interpreting and responding to distressing mental experiences, reduce ritualizing and avoidant coping, and improve functioning and quality of life. (You'll want to adapt these generic CBT goals according to which CBT approach you use; see subsequent chapters.) Stress that "CBT tends to work *when you do it*—how much you gain from treatment depends on how much you invest." The number of sessions recommended (excluding the initial assessment) varies according to the level of severity; my rule of thumb is as follows: mild: up to ten sessions, moderate: eleven to sixteen sessions, and severe: more than sixteen sessions.

At this point you have completed the initial assessment, established that the person has OCD, and provided treatment recommendations. The following chapters will separately consider CBT treatment options, starting with the flagship older approaches (ERP and CT) and moving on to more recent methods (IBT, MCT, and ACT).

CHAPTER 3

Exposure and Response Prevention Therapy (ERP Therapy)

ERP was briefly described in chapter 1. I am purposefully ranking it first in the lineup of treatments because it is the oldest, best-validated psychological therapy for OCD. Any therapist treating clients with OCD should ideally have a grounding in orthodox ERP, as this forms the basis for later developments in treatment. This chapter will start with a description of the theoretical model, then consider assessment and treatment. The primary reference text is the excellent treatment manual by Foa, Yadin, and Lichner (2012), of which some of the methods have been adapted below and integrated with other authors' suggestions.

Clinical Model

In chapter 1, the behavioral model was described in its historic context (see Mowrer, 1960; Rachman & Hodgson, 1980). Briefly, people's obsessional fears develop through *classical conditioning*, whereby the conditioned stimulus (thoughts, objects, or situations) acquires the ability to trigger anxiety or discomfort because of its association with an unpleasant stimulus (the unconditioned stimulus). After experiencing the obsession, they may perform a compulsion or use another strategy for reducing their anxiety or discomfort, reassuring themselves, or preventing feared consequences from occurring. The compulsion or safety strategy is then maintained through a process of *negative reinforcement*; that is, it serves as future encouragement of behavior that is experienced as being effective in reducing negative feelings or preventing feared consequences (see figure 3.1).

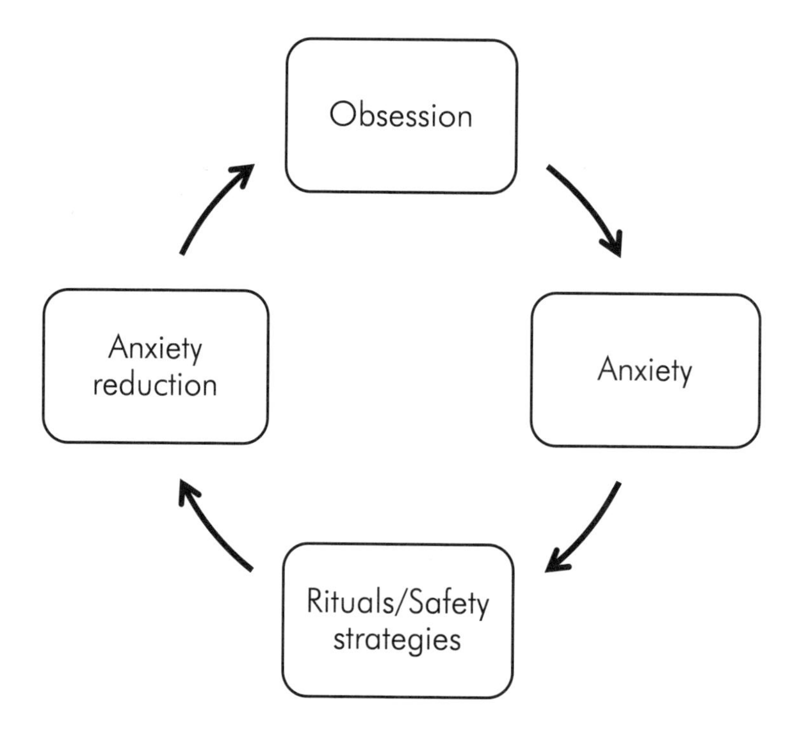

Figure 3.1. The behavioral model of OCD

However, anxiety reduction is only temporary, and the compulsion has to be repeated the next time the person experiences an obsession. In this way, the obsession-compulsion cycle is established because performing compulsions sabotages the durable reduction of fear, which requires extended exposure to the obsession *without* the compulsion being performed.

Strategies for reducing or avoiding anxiety include overt or covert (behavioral or mental) ritualizing (e.g., washing, checking, mentally undoing a thought, ritualistically referring to "my horrible, unwanted thoughts" for self-reassurance when describing obsessions to the therapist), avoidance (e.g., not touching a "contaminated" item properly, thought suppression), or reassurance-seeking from others (e.g., asking, "Do you think I acted abusively?" without this being a realistic concern). Reassurance-seeking, which can be subtle, can also manifest in the session—a therapist may have a sense of getting squeezed into a corner when a client is suddenly, unusually persistent in a line of questioning.

The key functional distinction is between obsessions, which activate discomfort or distress, and rituals or safety strategies, which are aimed at reducing discomfort or distress.

Although the behavioral theory of OCD has limitations (see chapter 1), it still represents a powerful, relatively simple working model for intervention;

therefore, it is a good starting point for trainee therapists. Next, I will consider the nuts and bolts of treatment.

ERP Procedure

Modern ERP retains aspects of the procedure first described by Meyer (1966). First, a hierarchy of obsession triggers (feared situations) is constructed, ranging from less to more anxiety provoking. Then, usually starting with a relatively easier situation, the client exposes himself to the obsession trigger and the accompanying anxiety (*exposure*) while not performing ritualizing or using any other safety strategies—at least until his anxiety or distress has decreased by a significant degree (*response prevention*). Exposure can take place in real-life situations (in vivo), or it can be imaginal, involving exposure to feared thoughts, such as about the feared consequences of not performing rituals, or about being in feared situations. Before discussing the steps in treatment, I will consider how ERP causes symptoms to change and which features of treatment are associated with success.

Treatment Mechanisms

Emotional processing theory (Foa & Kozak, 1986) stresses the importance of a process of *habituation* (or "numbing out"). This occurs when there is repeated, uninterrupted exposure, of a sufficient duration, to feared stimuli. Anxiety escalates, peaks, plateaus, and eventually declines (or habituates) during extended engagement with the stimulus. Compulsions, with the intention of ensuring safety and reducing anxiety, interrupt or sabotage this process.

However, some recent research has questioned some of the assumptions of emotional processing theory (see Baker et al., 2010). For example, successful habituation during exposure is sometimes not associated with long-term benefit. Psychologist Michelle Craske and colleagues theorized that *inhibitory learning* is more important in allowing durable extinction (i.e., reduction or elimination) of anxiety (see Craske, Treanor, Conway, Zbozineck, & Vervliet, 2014).

In a nutshell, according to inhibitory learning theory, the original anxiety-elevating association between the unconditioned and conditioned stimulus (i.e., processing that there is a threat, for instance that door handles may be contaminated with dangerous germs) is left *intact* during extinction. The aim of treatment is to create new nonthreat associations (door handles are generally safe), which compete with the original threat association. Therefore, the goal of ERP should be to maximize the creation of new nonthreat associations to inhibit activation of the old threat associations (i.e., inhibitory learning); the emphasis is on maximizing the durability and generalization of learning during treatment so that after treatment, the new nonthreat associations override and supersede the old threat associations, across settings.

ERP theorists also recognize the contribution of cognitive change. This occurs during and after exposure when clients learn that they can manage their anxiety, which tends to decrease, and that their worst fears are not realized. There is also a likely increase in *self-efficacy* (the belief in one's capability to achieve a goal; Bandura, 1982) when clients have the experience of mastering their OCD. In ERP, exposure is approached as hypothesis testing of the client's specific idiosyncratic predictions about feared emotional or actual consequences (Foa et al., 2012).

Factors Associated with Therapy Effectiveness

ERP benefits from a lengthy track record of investigation through practice and research. Various factors have been identified with relevance to how well treatment works (see Foa et al., 2012):

Duration of exposure. Prolonged, uninterrupted exposure has been found to be more effective than the alternative. A commonly stated rule of thumb is to aim for a ninety-minute session, but sessions can be briefer or longer depending on the client's progress in mastering the situation and achieving habituation.

Treatment scheduling. ERP can be administered in an intensive format, involving daily sessions over several weeks, or administered twice weekly or weekly. When making a decision, consider client preference, the severity of symptoms (severe presentations tend to require longer and more frequent input), and the client's ability to do homework exposure assignments (better adherence would allow less frequent appointments). Of course, realistically, parameters are also set by practical considerations, such as financial or service restrictions, the client's travel arrangements, work schedule, and so on. Particularly when exposure work with the therapist is more infrequent, daily homework exposure practice is optimal.

You can also supplement the client's clinic appointments with brief telephone contact or video conferencing, which can be useful for troubleshooting to allow self-directed work to proceed at an optimum pace.

Twelve to twenty treatment sessions, lasting sixty to ninety minutes, would suffice for the majority of clients with OCD. Several follow-up sessions, at three-month and six-month intervals, are recommended for ensuring maintenance of treatment gains.

Predictability and therapist style. A predictable and collaboratively planned exposure schedule ensures that clients know what to expect of exposure sessions and allows them the final say over the pace and content of work. This ensures that clients' motivation and compliance are not needlessly put at risk, such as if they were unexpectedly faced with a challenge that they were unable or unwilling to tolerate.

As a therapist, your role is to encourage, and sometimes (firmly) prompt and challenge. However, this should be in the context of ongoing monitoring and assessment of the client's progress and resolve, to maximize the likelihood of experiencing success. If more rapid progress in a session than was previously planned for seems possible, deciding to advance to a higher level of exposure challenge should be carefully, collaboratively considered with the client before proceeding. In earlier years I discovered the pitfalls of moving too fast in a session, with the planned schedule left behind in the dust. Apparently speedy progress, in a buoyant mood, may turn out to be too much, too soon. Despite mastery in the session, at later moments of uncertainty the client may appraise herself as having been lucky to have escaped unharmed. She may worry about what she will have to face at the next session, and this may put her therapy engagement at risk.

Finally, humor, judiciously used, can make a useful contribution. The radical cognitive shift in perspective of one's predicament, encapsulated within a lighthearted comment, can lighten the load sufficiently to make possible the critical engagement with exposure challenges. However, if the therapist is not comfortable in using this strategy skillfully, it is best avoided.

Step-wise ERP versus starting at the top of the hierarchy of difficulty. You are faced with a choice at the start of therapy: to begin by having the client tackle the most difficult obsession trigger, or to start at a lower, easier level in the hierarchy and administer exposure using a graded format. These approaches are equivalent in efficacy; however, clients tend to prefer a graded format. Given the priority of maintaining client motivation to comply with a difficult therapy, currently the most commonly used format is graded exposure.

Individual or group ERP. There is evidence that both formats can be effective. ERP is a structured therapy and easier to systematize than cognitive therapy (chapter 4), therefore lending itself well to a group format. Group work may be facilitated by selecting participants with overlapping symptoms, although this may not be as straightforward as it sounds, as many clients have obsessions in several domains. Setting up group work presents a greater logistical challenge, but participants can benefit from reciprocal support with other group members and getting fresh perspectives on their own problems by observing and participating in the treatment of others.

Therapist-guided versus self-guided exposure. Abramowitz (1996) found therapist-guided exposure to be more effective than self-guided exposure in alleviating OCD symptoms (some of the benefits may include the therapist ensuring an adequate duration of exposure and helping to maintain focus on the anxiety-provoking situation). Standard protocols include therapist-guided input during sessions, intermixed with self-controlled exposure as part of therapy homework.

ERP for OCD

ERP can be divided into an assessment and treatment phase, although assessment continues throughout as the complex process of treatment unfolds (which in my experience never goes exactly to plan!). Schematically it can be represented as follows.

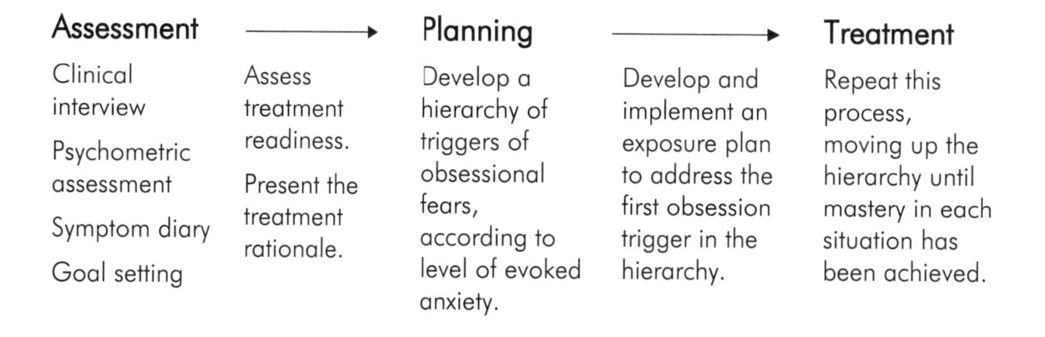

Figure 3.2. ERP: Stages in assessment and treatment

Assessment for ERP and Goal Setting

To conduct a thorough assessment, gather detailed information about (1) external triggers of obsessional fears (objects, places, situations), (2) obsessional thoughts that evoke anxiety (images, doubts, or urges that the person finds repulsive or unacceptable), (3) the immediate emotional consequences of encountering OCD triggers (anxiety, unease, disgust, guilt, shame, anger, sadness), (4) shorter-term and longer-term feared consequences if the trigger were encountered and rituals not performed, (5) behavioral and mental rituals (for reduction of anxiety or discomfort) and their scope and extent, and (6) avoided situations. (See chapter 2 for assessment questions.)

The data informing the case conceptualization from a behavioral model (figure 3.1) come from four sources: (1) the clinical interview, (2) self-report and clinician-administered scales, (3) symptom diaries, and (4) interviews with significant others (chapter 2).

If multiple OCD symptoms manifest in the client's home environment, a home visit may yield useful additional information, as it allows direct observation of obsession triggers, ritualizing, and avoidant coping, potentially enabling undisclosed symptoms to be identified.

Ask the client to identify shorter-term (therapy) and longer-term (life) goals (see Pollard, 2007), which she can specify on a recording sheet that can also be used for monitoring progress. One client identified these goals:

Therapy Goals

shop without checking my bag for concealed items before I leave the shop

get ready more quickly in the morning

stop nagging my family to do pointless things

stop retracing my route after driving

Life Goals

do more exercise

spend less time on Facebook

spend more time reading novels

Difficulty in identifying therapy goals may point to motivational problems, or these may emerge at a later point. Therefore, it's important to assess treatment readiness and develop strategies for addressing problems in this area.

Is the Client Ready for Therapy?

High rates of treatment refusal and dropout are important obstructions to ERP, so be on the alert for problems such as the client's not being ready for or not being motivated to engage fully with treatment. Most of the research in this area has been conducted with specific reference to ERP, which is why this section is included in chapter 3. However, there is clearly scope for application in the other CBT approaches as well.

Pollard (2007) lists four questions (adapted here) to consider: (1) Does the client understand the treatment model? (2) Are the client's expectations of treatment realistic? (3) Is the client adequately motivated? (4) Are there any apparent obstacles to proceeding with treatment?

Each of these can be assessed with simple questions to the client, which can be supplemented with questionnaires such as the Treatment Ambivalence Questionnaire (Rowa et al., 2014; see chapter 2) to allow a more rigorous assessment. Possible treatment-interfering behaviors can be identified already at the assessment stage, such as the appointment being arranged by someone other than the client, the client's unwillingness to complete questionnaires, poor compliance with the administrative requirements of the practice, a dismissive attitude, and poorly informed criticism of psychological approaches. These potentially point to a bleak prospect for therapy outcome and will have to be addressed.

To optimize the chance of successful client engagement with therapy, Pollard (2007) suggests the following: provide accurate information about OCD and ERP in a way that minimizes guilt ("Your symptoms are features of a recognized mental disorder, not a sign of inner weakness or deficiency; you are not to blame for them"), reasonably create an expectation of treatment benefit, and accurately outline the client's and your responsibilities ("You are the expert on your life and your symptoms, and I am the expert on how to help you reduce those symptoms").

It is important to have a frank discussion with clients about what will be required of them over the course of therapy, including regular homework exposure practice and completing worksheets. You can sidestep the possible negative connotations of "homework" by saying that together you will decide on an "action plan" for daily practice at the end of every session, to make sure that the benefits of session work generalize to the rest of the person's life. Ask clients how much time they could set aside for therapy work over the week and how they could temporarily free up space, thus reinforcing the view that therapy is a serious project and will require significant time and effort.

This may be a truism, but the information you provide should be research validated, thus allowing your credibility not to be damaged if your client challenges you on a point and you turn out to have had the wrong end of the stick! People with OCD (who, based on the client group served by my practice, often have research expertise and online access to scientific data) may spend a lot of time reading about OCD and be aware of findings the therapist is ignorant of. When stumped by a question, it is better to admit that you are clueless and will look into it rather than pretend you have more knowledge than you do.

Maltby and Tolin (2005) describe interventions that could be useful (but may present a practical challenge): the client could have a confidential conversation with someone (preferably with matched OCD symptoms) who has benefited from ERP, or, after all relevant consent has been obtained, could watch a video recording of ERP applied in session. You can also lead clients to other useful information and personal accounts of treatment success that can be found on reputable websites (such as the International OCD Foundation website).

Unfortunately, despite adhering to the guidelines listed above, you may find that the client is still ambivalent and reluctant to engage with treatment, or motivational problems may emerge later in therapy. In that case, it may be worth considering the use of specialized methods from motivational interviewing, described at the end of this chapter. Nevertheless, if despite your best attempts, poor treatment engagement and motivational difficulties persist, it may be better to delay starting therapy until the client is better equipped to make use of it, rather than risk exposing the client to the potentially demoralizing experience of treatment failure.

Presenting the Treatment Rationale

The following extract demonstrates the main ideas that need to be conveyed when presenting the treatment rationale to a client:

> *Okay, let's summarize the main parts that make up your OCD. The first part is your obsessive thoughts, which activate your fears or concerns; they cause anxiety or distress when you encounter a trigger. Understandably, because anxiety is unpleasant and you worry about what may happen next, you try to reduce your fear, and this is where the second part of the OCD—compulsive ritualizing—comes into play. Ritualizing reduces your anxiety and apparently keeps you safe, and because of this, repeated ritualizing becomes a persisting habit. Of course, another way of managing your fears is to avoid obsession triggers altogether, but this narrows your life and tends to hold you back.*
>
> *Let's illustrate this with an example: you mentioned that when you go grocery shopping, you end up worrying that you or someone else might have put something into your pockets or your handbag, and that's why you have to check thoroughly before you leave the shop. Here, the obsessive thought is that there is an unintended, concealed item in your pockets or bag. The obsession causes you to feel anxious because you worry you could be arrested for shoplifting. You then check your pockets and handbag extensively to reduce your anxiety and reassure yourself—this is the ritual or compulsion.*
>
> *This sequence has become a pattern, and now every time you find yourself in a shop, you experience similar fears and find it difficult to resist checking even though you may know at a rational level that there is no realistic*

basis for your fears. Over time, checking ends up evolving into an entrenched habit. And even though it reduces negative feelings, can you name a few problems with this solution? (Let the client list the disadvantages.) *What also happens is that you don't realize that if you didn't check, your anxiety would fade over time, you would find you can cope with anxiety and uncertainty, and nothing bad would happen. By not checking, the link between the worry thoughts and anxiety is eventually broken, a new rational habit becomes established, and this overrides the concerns of the past.*

This is what we will try to accomplish in treatment, using a careful, systematic approach. We will make a list of triggers of your obsessional fears and arrange them into a hierarchy of difficulty. We will pick a situation that you can cope with, and I will help you to face the situation while not doing rituals or blocking your anxiety in any way. You will find that you learn to cope with the anxiety, that it reduces over time, and that your worst fears don't come true. We may use different exposure methods; sometimes exposure will be applied to actual situations and at other times to your thoughts about situations, such as troubling mental scenarios about what might go wrong if you didn't perform the ritual. We will work our way up the trigger hierarchy until you have conquered them all. It will be important for you regularly to set time aside for exposure work, in addition to what we do in sessions. We will aim to completely eliminate all OCD symptoms, which is much better than trying to negotiate with the OCD. It's like the surgeon removing all of the cancer rather than just 90%. It will be important for you to work on staying engaged with the full course of therapy, and I will support you in accomplishing this.

It may seem like a mountain to climb at this point, but the far majority of people who do this treatment find their confidence increases substantially when they get going with the program and realize their fears are unfounded. But don't take my word for it; let's find out. Is there anything you don't understand? (This can be supplemented with: *Because it is so important that the treatment approach makes sense to you, can you tell me, in your own words, how you would explain exposure therapy to a friend this evening? I don't want to put you on the spot—just try and give your friend a rough idea.*)

Constructing a Hierarchy of Obsession Triggers

This hierarchy, developed in collaboration with the client, is informed by the data-gathering strategies listed above and provides the framework that will guide the breadth, detail, and sequencing of exposure work. As previously described, a meaningful division is between external triggers (situations, objects, people) and internal triggers (feared thoughts, images, memories, imagined catastrophic scenarios) of obsessional fears. The category of internal triggers, such as having an unprompted thought about possible harm to loved ones, applies where the activation of anxiety is less closely and reliably linked to a specific, identifiable external stimulus, such as seeing an attractive person that triggers an obsession that one does not really love one's spouse.

With the client, draw up a list of between approximately eight and fifteen situations that closely reflect the client's fears. When clients present with symptoms in multiple domains (e.g., contamination, checking, aggressive), it is simpler to draw up hierarchies for each domain separately. It is usually impossible or impractical to list every possible trigger of the client's fears, but the list should ideally cover a range of difficulty, including the client's "worst fears," and include the areas where OCD-related impediments are most visible and cause the most frustration.

Describe hierarchy items in sufficient detail (e.g., park benches, cinema seats, eating with friends, eating in restaurants, using the office restroom, using a public restroom) to allow sufficiently structured planning. But allow enough scope within each item or category for a flexible approach to include exposure in different settings, which may represent different levels of difficulty, for maximizing inhibitory learning. For example, in the category of public restrooms, the list might include using restrooms in several different malls, varying according to cleanliness and the level of usage, which may vary according to time of day (see "Treatment Mechanisms" above for a discussion of inhibitory learning).

After drawing up a list of triggers, the next step is to rate the items according to the subjective units of distress (SUDS) on a scale from 0 to 100 (none to maximal distress) that they evoke ("How much discomfort would you experience if you..." [therapist describes the exposure] "without performing..." [therapist describes safety strategies used]). The use of SUDS has a long track record in behavior therapy, but I prefer the simpler terminology of assigning an anxiety/discomfort/distress rating from 0 to 100 (select the label to provide the most accurate description of the client's emotional response, or according to the client's preference). The following verbal descriptions for different scale points can be helpful: 0 = no anxiety, 0–25 = mild anxiety, 25–50 = moderate anxiety, 50–75 = high anxiety, 75–100 = severe anxiety, 100 = maximum anxiety (which can be defined as how you would feel if you were trying to swim to shore while being circled by a great white shark!). See an example of a trigger hierarchy below:

Germ and dirt contamination	Discomfort (0–100)
Public restroom toilet	100
Floor in a public restroom	95
Toilet in office building	80
Floor of restroom of my office building	75
Doctor's reception area	75
Shaking hands with a stranger	75
Train seat	65
Bus seat	65
Door knobs in public building	60
Door knobs in my office building	55
Someone else's telephone at work	55
Someone else's computer at work	50
Someone else's chair at work	30

Deciding Where in the Hierarchy to Start Exposure and What Defines the Endpoint

The traditional ERP protocol advises starting exposure at a point in the hierarchy that is neither too easy nor too difficult (on average, discomfort at about 40–50, although higher or lower depending on the client's ability to tolerate anxiety), thus allowing a good chance of the client's experiencing noticeable habituation and successful mastery of the situation.

However, Abramowitz and Jacoby (2015) caution that this progression validates a problematic assumption that higher anxiety *objectively* represents higher threat or challenge. Ideally, the structure of therapy should strengthen a realistic assumption. Therefore, an alternative approach is to start with situations presenting a higher level of interference in the client's functioning, or simply to use random selection (this strategy is also supported by contentions derived from the inhibitory learning model; see Craske et al., 2014). Depending on whether this is practically feasible, several lower- or intermediate-difficulty situations can be selected as a starting point and addressed alongside each other.

You might have to be innovative in identifying lower-difficulty situations when the client has rated *all* situations as being highly distressing. For example, if the client has rated writing checks as 80 (for fear of making a mistake), you may prompt him to consider whether writing a check for your reception staff will be easier, thereby allowing a lower rating.

Hereafter, when following the conventional approach, exposure progresses stepwise from lesser to more anxiety-provoking situations, using different exposure methods (described below). The aim of therapy is for the client to attain mastery in all the fear-inducing situations, meaning that when encountering the trigger, the client's anxiety is manageable and he can respond flexibly and without needing to resort to maladaptive coping strategies.

Treatment Settings

It may not be possible to access trigger situations in the clinic setting. Therefore, to conduct supervised exposure, it may be advisable to arrange to meet the client at her home (for example, when working on compulsive cleaning or checking, where the client's fears are linked closely with her domain of responsibility—her home), or in a public setting (for example, when working on fears about exposure to human sweat in a public exercise area). If so, this needs to be with the client's full consent, and steps should be taken to minimize avoidable encroachment on her privacy. If meeting in a public setting, the detail of exercises should be predecided and privacy issues considered; confine discussion to a relatively private area so it is inaudible to members of the public.

ERP Strategies

Different strategies are available in ERP depending on the nature of the anxiety-eliciting stimulus.

IN VIVO EXPOSURE

In vivo exposure is the natural approach used for overcoming fears such as attending primary school for the first time despite qualms about not fitting in, or having driving lessons despite worrying about crashing the car. It requires facing obsessional triggers in real life, without preforming rituals. It generally accounts for the bulk of exposure work. Below are examples of in vivo exposure:

- After touching a red spot on the clinic carpet (obsession: *Could it be infected blood?*), the client refrains from washing his hands, using antibacterial wipes, or engaging in lengthy mental calculations about the risk of infection.
- The client touches a tissue on a park bench used by drug dealers and wipes it on her face, therefore contaminating herself with a character-corrupting moral pollutant (tissues are quite handy props for transferring "contaminants" to the clinic setting for exposure work).
- The client does not check the gas cooker when leaving the house.
- The client purposefully puts objects on his desk in the "wrong" place.
- The client crosses the road intentionally to walk past "vulnerable" people when fearing that she might attack them, while holding in mind thoughts of attacking them.
- The client intentionally floods his mind with swear words and blasphemous thoughts when attending a church service.

Next, I will separately outline the steps required to administer in vivo exposure.

Develop a situation-specific exposure plan. At this point, a trigger situation (or several) has been selected, and your focus should be on introducing a sufficient level of structure and support to maximize the chance of success. It is useful, in collaboration with the client, to specify a *behavioral goal* to be attained in the situation. It may not be unreasonable to define this conservatively as how a person *without* OCD would act in the situation (e.g., use the public restroom, briefly wash hands, use the exit door handle as normal, handle personal possessions as needed). However, if such a goal were attained successfully, it may not be readily apparent that this may leave the client restricted in his behavioral repertoire and vulnerable to becoming unhinged when unforeseeable developments occur (e.g., dropping his handkerchief in the restroom). Such events may pose novel challenges by activating obsessional fears, thereby exposing the client's underlying vulnerability, and leading to symptom escalation.

Therefore, it is better to define the behavioral goal ambitiously ("going the extra mile") as comprising the *full* range of responses accessible to a reasonable and responsible person without OCD, which the person would consider as within the spectrum of acceptable risk. For example, most people (including me!) do not like sitting on the floor of a public restroom but would be able to do so without experiencing overwhelming anxiety or disgust. If this were suggested as a behavioral goal (sit on the floor for one minute, touch hands to the floor, then handle personal items without washing), the client might object: "If nobody does this as part of normal life, why should I?!"...to which the therapist can respond that these are the measures required for vigorous, successful treatment of OCD. People who do not have headaches do not take ibuprofen; being able to "go the extra mile" increases the client's freedom and flexibility and holds a benefit also for other items in the hierarchy, higher and lower, by demonstrating that apparently challenging situations need not be feared and can be mastered using a focused approach.

It may be reasonable for the client to consider what constitutes "normal" behavior in the situation, but this question may be more complicated than it seems. Years of ritualizing may have left the client with a lived-in, distorted view of what defines reasonable behavior, with little sense of any alternative. Ask the client to reflect on and be guided by her premorbid behavior or her observations of trusted others' behavior, which could be aided by surveying her friends and family, asking specific questions, and collating the information. Suggest that the client "step into [her] non-OCD, rational best," reflect on the facts at hand, and make a decision. It may be helpful to offer your opinion or describe your own behavior in the feared situation, but over the course of therapy, momentum should build for the client to take the initiative in charting the course of rational action.

It is also worth considering the recommendation by Craske and colleagues (2014) to maximize the opportunity for *expectancy violation* in designing exposures: the more the client's expectation can be violated by the outcome of exposure, the greater the inhibitory learning (see above). This view overlaps with models in which exposure experiments are used for belief disconfirmation (see chapter 4). Therefore, also consider the extent to which attainment of the behavioral goal would offer the most thorough disconfirmation of the client's fearful expectations, so as to maximize inhibitory learning. This rationale also vindicates "going the extra mile" goals (see above).

After clarifying the behavioral goal, the next step is to consider whether to pursue exposures as a single step (preferable) or a series of steps (graded); that is, expose and eliminate all safety behaviors in a single step, or break down attainment of the behavioral goal into several steps in a mini-hierarchy applicable to that specific trigger.

Grading can be accomplished by modifying:

- The duration of exposure (e.g., incrementally increased duration of use of a public computer, without first wiping the keyboard, while intentionally handling personal items)

- The interval before rituals are performed (e.g., an increasing time interval separating use of the computer and the first hand wash)
- The duration or number of rituals (e.g., reducing the duration of washing, or instead of performing an action in three sets of four, performing it in two sets of three)
- The ritualizing method (e.g., using normal soap instead of antibacterial soap and eventually no soap at all, or reversing the sequence of ritualistic checks)
- The amount of reassurance given by others (e.g., saying a single word instead of a phrase)
- The presence of the therapist or a support person at the time of exposure (e.g., initially someone is present, but eventually the client completes the exercise independently)

The need for grading may emerge after failure in attempting single-step exposure, despite the therapist's best efforts. However, ideally the planning of exposures should be focused on presenting the client with the optimal level of challenge: that which will bolster motivation by allowing maximum improvement but not exceed the client's coping ability, leading to erosion of self-efficacy through failure.

Execute the exposure plan. Firmly encourage (but never coerce) clients to engage fully with exposure and work their way toward attainment of the behavioral goal, either as a single step or as part of a graded strategy. For example, after firmly and comprehensively touching "contaminated" objects, the person may be instructed to thoroughly touch his face, hair, or personal belongings (e.g., cell phone, wallet), or touch a finger to his mouth or tongue, *without* any subsequent ritualistic cleaning or any other safety strategy (or as agreed as part of a graded strategy). This sequence can be repeated multiple times over the session. (An Exposure Recording Form is provided in appendix 1 to assist in record keeping; this can also be downloaded at http://www.newharbinger.com/38952).

Advise clients that they should let themselves be guided by the aim of "seeking out the anxiety" rather than find ways of making it easier. Rather than focusing on safety, they should apprehensively contemplate disasters and feared consequences (thus, optimizing new learning by combining in vivo with imaginal exposure—see below). Consider the recommendation by Craske et al. (2014) here for "deepened extinction": conducting exposure to fear-inducing stimuli separately (e.g., first, exposing the client to thoughts about attacking his child, and second, to a situation where he is holding his sleeping child) before conducting combined exposure (doing both at the same time).

It is good practice, in the best spirit of the scientist-practitioner model, to regularly request anxiety/discomfort ratings (0–100) over the course of exposure, perhaps every ten minutes or so. This requires the client to focus on his anxiety (which may facilitate habituation) and allows you to gain insight into the client's experience using a more rigorous quantitative criterion and a more exact assessment of whether habituation has been achieved. Anxiety ratings can eventually be graphed and presented to the client as powerful, morale-enhancing visual confirmation of his progress.

Your questioning about exposure should seek to elucidate and describe the client's experience: "What are your feelings/fears? How are you feeling in your body?" (Some research suggests that emotional labeling during exposure can be helpful for inhibitory learning; reviewed in Craske et al., 2014). Observations and interpretations that you offer should encourage (realistic) processing of how the client's anxiety level is improving, how he is managing to tolerate and accept his anxiety, and how he is coping with facing the situation without the burden of performing rituals.

Resist the intuitive urge to act to reduce the client's anxiety or provide reassurance that his fears are unfounded, and instead provide moral support and encouragement (e.g., "I know you can do it…"). If the client experiences very strong emotion (including anger or a panic attack) or disengages from exposure, take a step back and discuss it, but then encourage the client to reengage with the exposure, ideally until noticeable habituation has been achieved or sufficient data have been generated to foster a conclusion of safety. The unwaveringly reinforced message is that anxiety is a temporary, subjective experience that may be very unpleasant but is not dangerous, and it is best faced head on and not allowed to encroach on the client's freedom by restricting his choice; rational judgment should determine action.

However, in the event of overwhelming, incapacitating distress, in order to maintain the client's collaboration and compliance, it may be necessary to consider options for delaying exposure work until the next appointment or for a "treatment holiday" (also see "When to Terminate Therapy Prematurely" below).

It can be useful, particularly in the earlier phases of therapy, to model exposure (e.g., touching the bottom of your shoe and then eating a biscuit with the same hand) before the client is required to engage with the exercise. However, such modeling should gradually be withdrawn as you move into the later phases of therapy (or for later exposures during a single session), so as to encourage development of client self-efficacy and autonomy. A similar principle applies to the design of exposure exercises.

When doing exposure work, special consideration should be made to safety strategies, engaging in activities for passing time, the use of a support person, and homework. I'll discuss each of these below.

Safety strategies during exposure. During an exposure, the client may lapse into ritualizing, either behavioral (e.g., cleaning, checking, brushing off, shaking) or mental (e.g., repetitively telling herself that all will be okay, performing a mental calculation to change an unlucky number to a lucky one). Encourage her to "sabotage" the compulsion, such as by reexposing herself to the contaminant or undoing the ritual (e.g., by recontaminating her hands, rethinking the "evil" thought, imagining the worst consequence, reremembering the unlucky song).

Sensible-sounding thoughts may act as rituals if the nonoccurrence of feared consequences is unreasonably attributed to them. For example, a client might say, "I thought about stabbing my husband, and then I thought that this was just OCD and all will be okay." In this case it would be worth exploring the ritualistic potential of "this was just OCD and all will be okay"; preferably it should be put to the test by performing the exposure without any self-reassurance. It is also essential for clients not to suppress distress-causing obsessive thoughts, but rather entertain them deliberately. For example, you might say to a client, "Do not try to block or suppress the thoughts that you may have poisoned your husband; instead, actively bring them to mind without trying to reassure yourself in any way."

Be vigilant for any behavior during exposure that might seem atypical, excessive, or unexpected. This may be a marker of subtle avoidance or ritualizing (e.g., brushing, patting, or blowing off contaminants, or using hand cream instead of washing), which you should ask about and address if necessary.

Behavior that may be appropriate in an earlier stage of exposure may no longer be helpful at a later stage if it deviates from what would be considered normal, flexible, and appropriate. For example, reflecting on risks about exposure and debriefing with another person afterward should be considered as possible safety strategies that are eventually best eliminated.

A level of flexibility and impromptu suggestion can be useful when doing exposure work. An example from my own practice was when I assisted a client with exposure to personal possessions that had made contact with the floor, which he feared was contaminated with human urine. I noticed at the conclusion that the client returned all the possessions to his trouser pockets, which seemed quite full. On further enquiry, the client admitted that he was less concerned about returning the items to his trousers (which would eventually be washed) than about putting them into his leather bag (which would be more difficult to clean). He was willing, as a homework assignment, to repeat the exercises practiced in the session, but also extend the work to include putting the items in his leather bag.

Activities during exposure. While resisting performing safety strategies, it is legitimate for clients to engage with activities, such as tidying, reading a book, or watching television. The aim is *to pass time*, while reconciling themselves with the experience of having intermittent or ongoing obsessive thoughts (note: this is not framed as *distraction*, of which the ostensible aim is disengaging from the obsession). Alternatively, they can practice agreeing with the OCD (*Yes, I may get sick from not washing my hands, but I'm not doing it anyway*, or *I may completely lose my marbles if I don't tidy the room, but I'm leaving it a mess anyway*, or *Yes, I could cause an accident by having this thought, but I'm thinking it anyway—bring it on!*).

Use of a support person. It may be useful (particularly with more severe OCD) to involve a friend or relative to assist the client when doing exposure work as a home assignment (see below). Brief the person on how to provide moral support and encouragement, which may be necessary particularly prior to and after the exposure, when the client has to resist ritualizing. Preferably, the helper should attend at least one session with the therapist to discuss the forms of support that are helpful and those that are not. Examples of helpful support are encouragement to engage with exposure ("I know you can do it" or "If you stick with the therapy, you will start feeling stronger") or to not perform rituals ("The psychologist said it is really important for you to try to resist doing rituals; it will get easier if you keep at it") and prompting the client to engage with an alternative activity to pass time, rather than lapse into ritualizing (initiating a conversation about an unrelated topic, directing the client to watch TV or go for a walk).

An example of *unhelpful* support is reassurance provision ("Don't worry, I am sure nothing will happen" or "I didn't see anything wrong"); the guideline should be to not do anything that reduces the client's anxiety about ritual omission. Instead, the client and support person can agree in consultation with the therapist on how the support person can most helpfully respond when the client asks for reassurance. This can be challenging in that the client may anxiously be applying pressure at the time, or there may even be a benefit for the support person, such as if providing reassurance would allow the client to relax and participate more fully in social activities. Therefore, prenegotiating the support person's response if the client applied pressure can be useful for preventing angry escalation at the time of exposure exercises at home ("I know this is very difficult for you, but we agreed with the psychologist that for me to answer your questions would hold you back and that it would be best for me not to say anything").

Exposure homework. Following exposure work using any of the methods described, (preferably) daily home exposure assignments can be planned to allow continuation of session work or moving to the next step in the hierarchy. This also allows session work to generalize to other domains. As with in-session work, the aims are to allow habituation, learning, and mastery. Allow sufficient time at the end of the session for detailed planning, and at the start of the next session for review of homework. An Exposure Practice Form is included in appendix 1 to assist in homework planning and recording; it can also be downloaded at http://www.newharbinger.com/38952.

Process the exposure experience. Foa et al. (2012) specify that in their protocol, exposures (in vivo and imaginal) are approached as "hypothesis testing." Therefore, facilitating processing of the client's experience during and after exposure is an important therapy goal, to establish that

- anxiety reduces during exposure and continues to reduce with repeated exposures,
- distress can be managed without ritualizing or avoidance, and
- feared consequences do not occur or have a low probability of occurring.

Encourage processing by routinely asking the client, during and following exposure, what lessons have been learned. As OCD fears are quite idiosyncratic, it is important to establish exactly the content of the person's fears and her specific predictions and measure these against the data generated during exposure.

It is helpful to record descriptive and evaluative data in writing (e.g., what actually happened, framed as "what a video camera would have recorded" as well as anxiety ratings, interpretations, and lessons learned), to crystalize insight and enhance memory encoding and retention. An Exposure Recording Form, mentioned earlier, is included for this purpose (see appendix 1; also available for download at http://newharbinger.com/38952).

Here are examples of questions for reflection after exposure:

- What have you learned about what happens to your anxiety/distress/discomfort over the course of exposure?
- What happened to your anxiety/distress/discomfort with repeated exposures? What happened to your confidence?
- You feared the anxiety would overwhelm and disable you; so, what happened?
- You feared that the window might have been left open if you didn't check again, so how did things turn out? Is there a lesson here?

- You feared that you might get ill if you went and sat on a chair in the doctor's surgery waiting area. What happened? What does this mean? What have you learned? How does this relate to what you knew before your OCD got going?
- You worried that your sexual orientation would change if you allowed yourself to have sexual thoughts about someone of the same sex. You then did this and what happened to your sexual preference? What have you learned about the credibility of what OCD tells you?

The timescale of feared outcomes might vary from shorter- to longer-term. It is usually simpler to assess the validity of shorter-term outcomes—what will happen over the course of exposure, immediately after exposure, or in the days after exposure—as the data are readily accessible over time, allowing a powerful comparison with the prediction. Carefully record the timescale and ensure that the client attends to the outcome, allowing the new data to be processed and encoded—this requires the client to be able to articulate that his anxious prediction turned out to be inaccurate.

However, if the client's predictions reflected longer-term consequences (developing cancer in subsequent decades, or going to hell after death), which cannot realistically be assessed over the course of therapy, imaginal exposure is the appropriate strategy (or testing the prediction for a shorter time frame, anyway; see more in chapter 6). This will be considered next.

IMAGINAL EXPOSURE

An in vivo approach entails confronting real-life situations. However, where it is not possible, for practical or ethical reasons, to recreate the situations in real life, it can also be helpful to confront feared *thoughts* through *imaginal exposure*.

Imaginal exposure can readily be used in tandem with in vivo exposure, such as when the latter is used to expose a client to not checking repeatedly whether windows are closed when she leaves the house, and the former is used for exposure to an imagined scenario of the house being burgled (if the client offers a fearful account of such an event occurring).

Here are other examples of thought content that can be addressed using imaginal exposure:

- Thoughts of selecting the wrong partner and afterward suffering a relationship breakdown with catastrophic long-term consequences, ultimately ending up penniless and destitute
- Thoughts about having committed a sexual offense, being arrested, and bearing the burden of shame

- Thoughts about being responsible for a disaster, such as having caused a child's death through carelessness

The goals of imaginal exposure are closely aligned and overlap with those of in vivo exposure:

- To reduce anxiety and the negative emotional impact of distressing thoughts and images, thus reducing their potency and frequency
- To increase tolerance and acceptance of distressing thoughts and the associated negative affective states
- To allow sober and unflustered reflection on the improbability of feared consequences
- To encourage a distinction between thoughts and behavior so the client realizes that thinking distressing thoughts about actions or an event does not make them happen (this addresses the erroneous belief of "thought-action" fusion; see chapter 4)
- To provide an economical strategy for addressing the client's "core fear," as when a core fear underpins rituals performed in response to multiple triggers. For example, a client's common feared outcome of "a change of my moral character" might be triggered in response to situations such as shaking hands with an "immoral person," sitting in a taxi in which the previous occupant might have been a prostitute, and so on. In this case, imaginal exposure to thoughts about a catastrophic change of moral character might reduce anxiety and the need to perform rituals in response to several triggers.
- To supplement in vivo exposure, where this has been only partially successful because the client's core fear has been unaddressed. An example of this would be where the client experiences only partial anxiety reduction after repeated instances of abstinence from ritualistic checking whether doors and windows are locked when going to work in the morning, and this ongoing anxiety derives from lingering worries about the feared hypothetical outcome (in this case, being blamed for a burglary; see the example on page 74).

Steps in imaginal exposure. The first step in imaginal exposure is collaboratively developing a script with the client describing the client's feared outcome. The client should describe the feared outcome in the first person, as if it were all happening at that moment, across multiple sensory modalities, describing others' reactions and her own reactions—thoughts, feelings, and bodily sensations. The script should depict the client's worst imagined outcome, thus reflecting the progression of events to culminate in its final, dreaded conclusion. For example, if you create a script where the client is driving and knocks down a pedestrian without realizing it, be sure to include the person's other fears, such as the policeman knocking on the door, being convicted, being shunned by family, going to prison, and being brutalized by the other inmates (*if* these reflect the client's articulated fears).

Ask the client to provide as bold, compelling, and vivid a description as possible, and take heed to follow closely the natural unfolding of the client's narrative and avoid guessing outcomes or contributing elements that are not volunteered by the client. It may take some time, through questioning and exploration, to "flesh out" the description, asking questions such as "What happens then?" "How do you react?" "What are you thinking/feeling/hearing/seeing/smelling?" "How vividly do you see it from 0 to 100?" "What is your anxiety/discomfort from 0 to 100?"

Second, after having identified the main elements, make an audio recording of ideally the client, but otherwise you, narrating the script from start to finish; this should preferably be done in an animated way so as to convey the emotional tone. Some clients may express feeling uncomfortable doing this due to social inhibition ("I've never been much of an actress!"); in this case, it may help to assign the recording as a homework assignment, which can be kept private. Further exposure entails the client repeatedly listening to the recording with her eyes closed, uninterrupted, while imagining being in the situation as vividly as possible.

Depending on progress with habituation in session, listening to the recording can be given as a homework assignment. Foa et al. (2012) recommend twice-daily exposure, separated by a few hours. The duration of homework exposure can be roughly guided by the number of repetitions required to allow meaningful habituation; exposure can be discontinued when the anxiety peak during exposure has consistently been at a low level, say below 20, and when meaningful processing has been achieved.

Imaginal exposure can also proceed in a graded format, where the script and subsequent exposure to it proceed stepwise, with more difficult elements added incrementally. For example, if a client is more concerned about facing his family than about being arrested after a sex crime, the former scenario can be added at a later point.

As with in vivo exposure, it is important to coach the client to resist imagining doing rituals during and after listening to audio recordings (e.g., saying "God forbid this happens" as a way of eliminating risk and reducing anxiety) or avoiding exposure to particularly disquieting elements (e.g., by tuning out from "hotspots" while listening to the recording).

Following imaginal exposure, similar to in vivo, it benefits the client to process the "lessons learned," in other words, what the outcome of exposure has revealed about the inaccuracy of the client's fearful predictions about exposure (see "Process the exposure experience" above).

The following extract is an example of a script for imaginal exposure, in this case to the client's catastrophic thoughts about being responsible for the burglary of his house:

> *It's time to go to work. I leave the house and walk to the car. I ask myself if everything is locked. I don't feel confident, but I decide to leave it. I drive away, get to work, and start the day.*
>
> *My boss calls me over, looking disapproving, and says the police are on the line for me. I answer the phone. A policeman tells me there has been a burglary at my house and I should go there immediately.*
>
> *I feel sick to my stomach; my heart is pounding. I rush down, get into the car, and drive home. I can see the blue flashing lights as I approach the house. The police are parked outside. My wife is there, looking at me accusingly.*
>
> *A policeman tells me there was no sign of forced entry; the burglars entered through an open window.*
>
> *I feel sick to my stomach. I walk into the house. They took everything and the whole house has been turned upside down. Everything that was locked has been forced open. The safe is empty. They trashed my collection of model planes, which I've been collecting since I was five years old.*
>
> *I messed up again. I am devastated. This is all my fault.*
>
> *In weeks to come, the insurance people tell me they're not paying out. I'm facing ruin. My wife walks out on me. I live the rest of my life alone and rejected.*

When not to use imaginal exposure. Foa et al. (2012) offer the following guidelines for when imaginal exposure is contraindicated:

- Where in vivo exposure, which should always be the preferred method, is readily available as an alternative
- Where the client has difficulty in imagining the feared consequences by bringing a vivid image to mind
- Where the client does not articulate defined feared consequences associated with not performing a ritual

Keep in mind that people vary in their ability to picture scenarios, which is necessary for imaginal exposure to be effective.

Next, I will discuss the use of *looped recording*, which has a specific application to mental ritualizing.

LOOP PLAYBACK RECORDING

Loop tape (or digital) recording is particularly useful where the client mainly performs mental rituals in response to obsessions, such as blasphemous, immoral, or offensive thoughts; dangerous words; or unlucky words or numbers (Steketee, 1999). It is also beneficial where thought suppression is used as a coping strategy and where obsessional intrusive thoughts elicit high *anxiety* rather than other negative emotions (e.g., shame, guilt, or sadness—here, cognitive restructuring work can be particularly useful; see chapter 4). Mental rituals present a special challenge because the boundary between obsession and compulsion is far less clear-cut than for behavioral compulsions, and thinking is less amenable to control than behavior is.

The first step is for the client to audio record the obsession on a device that allows loop playback. The client will find the experience discomfiting but should gently, but firmly, be encouraged to do this after the rationale for the procedure has been explained to her:

Therapist: We will be exposing you to your obsession through listening to the recording over and over again while not performing rituals or trying to tune out in any way, so that you can learn that your anxiety diminishes of its own accord and that you can cope without needing to avoid the thoughts or perform mental rituals.

It may be necessary to clarify the distinction between the obsession and ritual to the client:

Therapist: Thinking of wanting your family to go to hell causes you anxiety; that's the obsession. You then say "null and void, this is not my desire, God protect them"; that's the compulsion, which reduces your anxiety.

This client could record, "I wish for every single member of my family—my brother, daughter, and wife—to burn in hell for all eternity," and explicitly omit the aforementioned compulsions (or perhaps add a compulsion-invalidating phrase: "and this is my sincere desire").

Hereafter, instruct the client to listen repeatedly to the recording on earphones, preferably in a quiet place that allows his full and uninterrupted attention to the content of the recording. Instruct him to allow the trajectory of his anxiety to escalate, plateau, and prevail until there has been a meaningful reduction in his anxiety (my rule of thumb is at least 20% reduction); this can be verified by having him record his anxiety every ten minutes (a cell phone app can be used to prompt recordings). A typical homework assignment may require twice-daily self-exposure sessions, around thirty minutes each, separated by an interval.

If there are multiple obsessions, a graded format can be introduced, administering exposure from easier to more difficult. Or, if a one-step approach is definitely not attainable, the content of a particularly distressing obsession can incrementally be framed in increasingly discomfiting terms, allowing stepwise exposure. However, there should be a firm resolution to tackle the most anxiety-provoking description as an end goal. "Going the extra mile" is also advisable here; in the example above, one option would be to visualize family members being thrown into hell.

When to Terminate Therapy Prematurely

Unfortunately, sometimes despite your best attempts at problem solving and negotiating obstacles (see previous sections), the client's progress continues to be stymied by treatment-interfering behaviors, such as persistent ritualizing or ritual substitution, noncompliance with homework assignments, and overt or covert avoidance. In this case, it may be advisable to terminate therapy prematurely ("live to fight another day"), so as not to expose the client to a demoralizing, protracted experience of failure. In this scenario, I would aim to ensure that the client has at least an adequate understanding of the principles of therapy. I would encourage her to contact me or another care provider when she feels ready to give her full participation in therapy.

Optional Module: Strategies from Motivational Interviewing During Assessment and Treatment

Some very interesting recent work investigates the application of the techniques of motivational interviewing in ERP for OCD (see Simpson, Zuckoff, Page, Franklin, & Foa, 2008). Here is a brief summary of the authors' work.

GENERAL CHARACTERISTICS OF MOTIVATIONAL INTERVIEWING (MI)

MI is a client-centered, goal-directed method for enhancing motivation to change (Miller & Rollnick, 2002). Drawing on Rogerian humanistic psychology, it is based on the view that human beings have an intrinsic drive toward beneficial change and self-actualization, but can be held back by ambivalence. It seeks to support the client's autonomy, facilitate collaboration, and evoke rather than instill motivation. Clients are approached as being rich resources, and their aspirations play a pivotal role in pursuing or not pursuing change.

The therapist's role is to help clients consider how change may help them achieve their goals and live consistent with their values, by empathetically asking open-ended questions and using reflective listening to communicate acceptance and deepen understanding. Clients' resistance to treatment is conceptualized as their experiencing ambivalence when put under pressure to take steps toward change. Rather than confront or direct, the therapist "rolls" with the resistance in order to resolve it. The therapist seeks to elicit talk about the importance of change and increase clients' self-efficacy, or confidence in their ability to change; thus, they benefit from being supported in articulating change-facilitating cognitions.

MI STRATEGIES DURING ERP

Simpson et al. (2008) conducted an OCD case series of ERP enhanced with MI techniques. The authors pursued all the standard ERP assessment goals but in addition added MI to evoke a commitment to treatment. Key features of the authors' approach as applied in the assessment and subsequent sessions (see the article for a full description) are described below.

Session 1: Present the introductory sessions to clients as helping them to decide whether ERP is right for them, rather than simply assuming it is. Elicit the client's perspectives on his symptoms and explore their impact, using open-ended questions and reflective listening. Handle the step of presenting a diagnosis sensitively as some clients may show resistance to the label. Evoke different vantage points on the desirability of change by asking the client to (1) predict the future if there were to be no change; (2) rate the importance of change, describe what makes change important, and consider what would be needed for the importance of change to increase; and (3) share his experience of past treatment and which treatment obstacles he encountered, and discuss his hopes for ERP, in order to evoke optimistic thoughts. In addition, when providing psychoeducation about OCD, use an "elicit-provide-elicit" format: *elicit* existing knowledge, *provide* new information after having obtained permission, *elicit* the client's reaction.

Session 2: Educate the client about ERP and develop an exposure hierarchy; this discussion follows the "elicit-provide-elicit" format described above. Also elicit positive and negative thinking about ERP and explore potential obstacles. To enhance self-efficacy, highlight the client's strengths and previous successes and engage in collaborative problem solving on how to circumvent and overcome obstacles.

Session 3: Reflect on the client's readiness for ERP and address ambivalence using the aforementioned strategies. Develop a change plan that summarizes the client's goals, reasons for wanting change, the ERP procedure, obstacles and their solutions, sources of support, and markers of progress for the client. Ask the client for a commitment to ERP.

From session 4 on, commence ERP, introducing MI strategies only when the usual methods of handling resistance are unsuccessful. Depending on the nature of the problem, you can evoke talk about the importance of exposure (e.g., the pros and cons of exposure, the anticipated gains, how exposure goal attainment would honor values), increase the client's confidence about doing the exposure (e.g., by reviewing past successes), or problem solve (ask the client which forms of support would be helpful).

When encountering ambivalence, ask open questions and reflect in an autonomy-affirming way rather than applying pressure and assuming the role of didactic expert aiming to persuade the client. The spirit of questioning is one of valuing and being genuinely interested in accessing the client's views and knowledge rather than using the Socratic method of leading the client to a therapist-predefined answer. Avoid confrontation—client choice and control are paramount.

The contribution of MI is sometimes more in the style than in the substance of strategy used (gearing down, temporarily, rather than changing cars); stylistic elements of MI are also common features of the collaborative application of CBT.

Here are some examples of therapist statements and questions (Simpson et al., 2008):

- You're finding this exposure difficult; what obstacles are you experiencing? (*exploring ambivalence*)
- This exposure seems an impossible challenge right now; can you think of a time when you succeeded with something that seemed impossible at first? How did you cope at the time? How can we help you? (*empathizing, increasing confidence, increasing "change talk," supportive problem solving*)
- How do you see your future if nothing changes? ("*looking forward*" *to bolster commitment to change*)
- Your rituals seem to help you in some ways, but in other ways they prevent you from being the person you want to be. ("*double-sided reflection*" *to explore ambivalence and highlight the discrepancy between maintaining the status quo and positive values*)

Relapse Prevention

If therapy exposure work has not been completed, make specific plans for how this is to be continued. Reflect on factors that previously triggered symptoms or precipitated relapse, and discuss with the client how he can respond to these using ERP methods, or how to access treatment if professional support is necessary in case of full relapse. Enquire about medication issues and give feedback to any referring medical practitioner.

At the conclusion of ERP therapy (and any CBT protocol), it is important to take stock of and conceptualize change. Therefore, with reference to specific OCD situations, ask the client to reflect on how his behavior, thinking, anxiety, and tolerance of the obsession have changed and review the key strategies that enabled change: (1) face the obsession or the feared situation; don't avoid it, and (2) stay in the situation long enough for anxiety to reduce and anxious predictions to be thoroughly tested. Fearful anticipation frequently makes the exposure more of a bogeyman than it turns out to be.

CHAPTER 4

Cognitive Therapy (CT)

Chapter 1 provided an introduction on how CT for OCD aims to address factors not addressed in ERP. In this chapter I will first discuss the cognitive-behavioral model of OCD and then outline the steps in treatment. To draw a clearer distinction with other cognitive-behavioral models, I'll use the more specific label of the cognitive appraisal model (CAM) and identify the associated therapy as *cognitive therapy*. However, readers should take note that this designates the more common description of the cognitive-behavioral model.

Clinical Model

The CAM of OCD derives from the generic cognitive model of emotional disorder, pioneered by the authors Albert Ellis and Aaron Beck, who, among others, made ground-breaking contributions. The cognitive model states that emotional disorder is caused by inaccurate, unrealistic, and unhelpful meanings attached to the self, others, situations, or events, which contribute to pathological emotional states and behavioral responses that maintain the disorder (see Beck, 2011). Further, Beck's (1976) cognitive content specificity hypothesis proposes that different cognitive content characterizes different psychological disorders, such as that the themes that characterize anxious thinking will reliably differ from those that characterize depressive thinking (danger and threat versus hopelessness and self-denigration).

Different levels of cognition can be distinguished: at the surface, automatic thoughts (including thoughts or images that just "pop up"); at the intermediate level, conditional assumptions (if A, then B) and personal rules; and at the deepest level, core beliefs, or basic assumptions (e.g., about the self, others, the future). Ongoing reciprocal feedback exists between cognition, emotion, physiology, and behavior; for example, negative mood both derives from and invites negative thinking.

How does this apply to OCD? The CAM builds on research findings that the far majority of ordinary people without OCD describe the experience of unwanted, intrusive thoughts, urges, or images (Purdon & Clark, 1993; Rachman & De Silva, 1978). People tend to experience intrusions during negative emotional states, and those who experience them try to resist them. The main difference between OCD obsessions and the intrusive thoughts experienced by nonclinical groups was not in their content, but rather that clinical obsessions lasted longer, were more frequent, caused greater distress, and were harder to dismiss (e.g., Morillo, Belloch, & García-Soriano, 2007; Rachman & De Silva, 1978; however, see some of the considerable complexities in this literature addressed in reviews by Berry & Lasky, 2012, and Julien, O'Connor, & Aardema, 2007). Ritualistic behaviors (e.g., checking, magical "protective" behaviors, washing) in nonclinical groups are also common (Muris, Merckelbach, & Clavan, 1997).

Therefore, the CAM states that there is a continuity between normal and abnormal obsessive intrusions. Intrusions can be seen as the product of an "idea generator," centered on the person's current concerns and interests (Salkovskis & Freeston, 2001), which functions as part of normal problem solving. Thoughts are subjected to further evaluation and will be experienced as compelling if further action is deemed necessary; alternatively, the thought may be considered irrelevant to current concerns and ignored.

So how do normal intrusive thoughts escalate into clinical obsessions? This depends mainly on two processes, which become key targets in treatment: (1) the *appraisal* of intrusive thoughts as being dangerous, abnormal, or harmful, requiring an action to be performed to prevent feared outcomes; and (2) the use of *safety/control strategies*: thought neutralization, compulsions, avoidance, thought suppression (described to me by one client as trying to hit the "snooze button" on his alarm clock), or reassurance-seeking from others. A key assumption is that the obsessive thought per se is not the problem; instead, the problem is the *meaning* that is attached to it (or the appraisal or interpretation of the thought), necessitating the use of control and safety strategies (see Salkovskis, 1999).

In the short run, the use of control strategies and ritualizing will result in a reduction in anxiety and negative emotions, an escape from doubt, and an increase in the sense of having control over the obsession; therefore, their future use is maintained through negative reinforcement (removal of an unpleasant experience). However, longer term, their use prevents people from learning that what they feared might happen if they simply left the situation unattended and did nothing doesn't occur; also, the nonoccurrence of negative events can be taken as supporting their effectiveness (they must help, because nothing happened). Further, their use may increase the frequency of intrusions because they serve as obsession reminders (e.g., where checking the tap triggers images of the house flooding) and because attempts at thought suppression might paradoxically result in an increase in thought frequency. Finally, hypervigilance for obsession-related stimuli results in an increased rate of detection (which is like seeing Fords everywhere when you want to buy a Ford). The end result is that the obsession increases in frequency, salience, and intensity, and the person seems less able to control it, attracting further negative interpretation and anxiety in a downward spiral. A simple diagram of the model is represented in figure 4.1.

Three levels of thinking can be distinguished in OCD: (1) intrusions, triggered by internal or external triggers, (2) appraisals, and (3) beliefs—assumptions held across situations.

In the CAM conceptualization, the unhelpful appraisal of the intrusion is the clinical kicking-off point for the obsessive-compulsive sequence and has been identified as being a productive therapy target. This has opened up new vistas for research and clinical innovation, leading to the identification of different kinds (or domains) of dysfunctional appraisal, which I will consider next.

Dysfunctional Appraisals in OCD

An international research collaboration (the Obsessive Compulsive Cognitions Working Group, or OCCWG; see Frost & Steketee, 2002) has put forward six themes represented in dysfunctional appraisals in OCD and the beliefs they derive from. In psychometric studies, the six themes cluster in three groupings of two each (OCCWG, 2005), described below.

CLUSTER ONE

Overestimation of threat. This refers to a tendency to overestimate the likelihood of danger, such as of making mistakes and believing they will have grave consequences. A person might believe *I'm an accident waiting to happen; if anything can go wrong, it will happen on my watch.* She may assume that certain situations are dangerous and that safety first has to be proven definitively, in contrast to an average person, who will tend to assume that things are okay if there is no clear evidence of danger. Given that in this OCD view, danger or error is probable and likely, guaranteeing safety may require extensive repetition of actions such as checking that the oven is turned off (because a house fire is highly likely and will be devastating because insurance won't pay out) or handwashing (because serious illness is highly likely if a spot is missed).

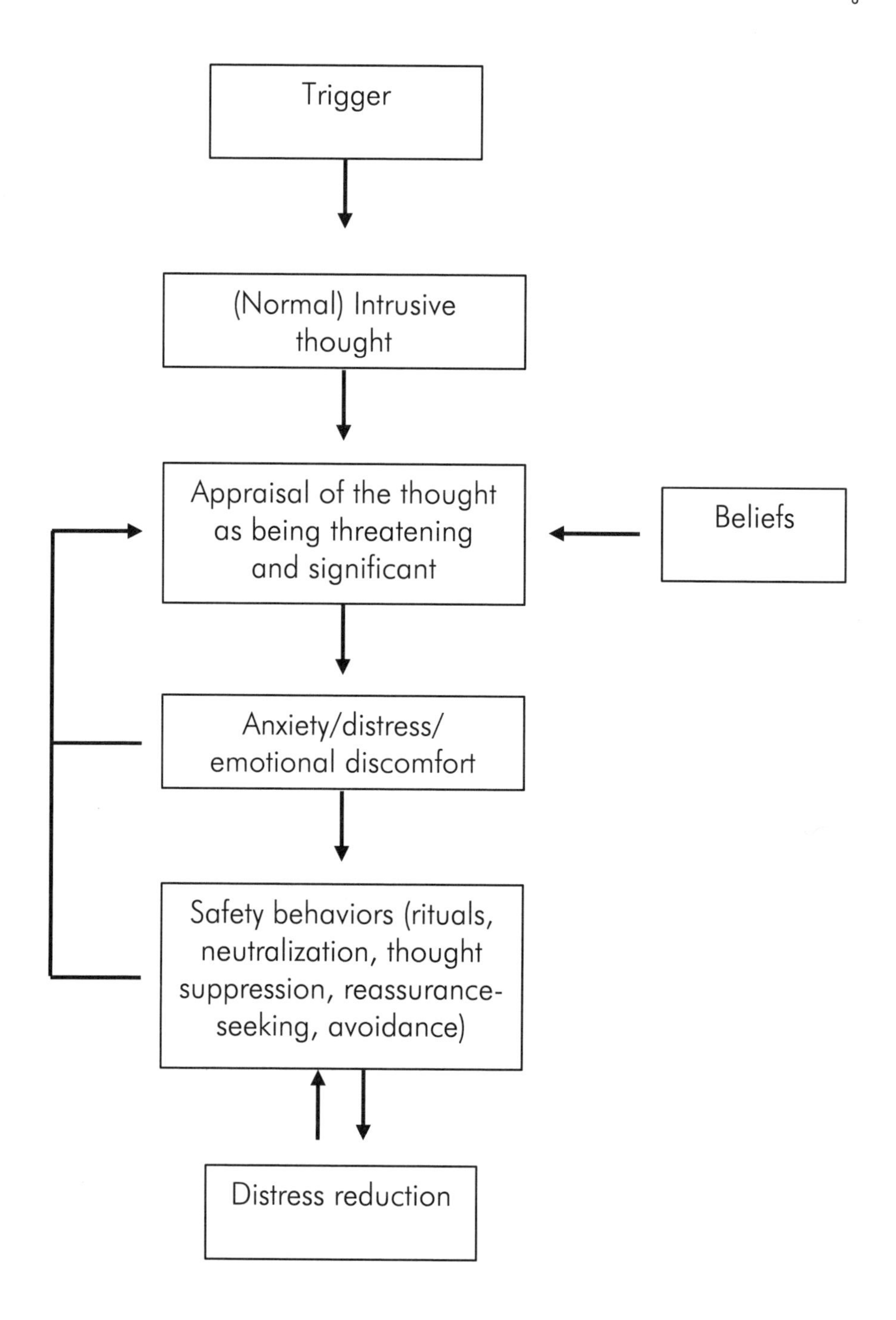

Figure 4.1. Cognitive appraisal model of OCD

Inflated responsibility. People with OCD tend to have a low threshold for assuming responsibility for certain negative outcomes and considering themselves culpable, which can go hand in hand with a tendency to overestimate the likelihood of danger (see above). Inflated responsibility has been defined as the belief that one has "pivotal power" to bring about or prevent personally relevant negative consequences (see Salkovskis & Forrester, 2002). These can reflect real events (e.g., causing a fire or harming someone by acting violently) or moral implications (e.g., being a bad person for having entertained "horrific" thoughts). Immorality or neglect of duty can be reflected in performing incriminating actions (e.g., shoplifting on impulse) or omitting necessary actions (e.g., not having pruned a shrub that partially obscured a traffic sign).

Examples of inflated responsibility thinking are:

- *Thinking something is as bad as doing it*—that is, unreasonably assuming responsibility and blame for having had a thought, which is confused with an *action*. This is also called moral thought-action fusion (the abbreviation, TAF, will be used from this point on). In OCD, TAF will tend to lead to the person's seeing it as his responsibility to use thought-control strategies to prevent feared outcomes.
- *It's as bad not to do something that may possibly prevent an extremely unlikely negative event as it is not to do something that will definitely prevent a very likely negative event.*
- *I'll be 100% to blame if...happened*—even if many other people's actions would also have contributed to the causal chain that resulted in the negative consequence.

Not surprisingly, inflated responsibility appraisals frequently concern the possibility of causing harm to *others*, resulting in anxiety being accompanied by guilt and shame.

CLUSTER TWO

Overimportance of thoughts. This reflects an exaggerated view of the significance of subjective experience. Thoughts of no consequence or little significance are appraised as being in a more significant category (i.e., as menacing, important, revelatory, or powerful), for example:

- A thought with violent content, in the absence of an intention, gets confused with a thought reflecting a real intention to act violently (also called likelihood TAF—*just because I think it, I'm more likely to act on it*). Or, a further example of this kind of TAF: a thought may erroneously be attributed causal properties—*if I think of my partner being in a car crash, it makes it more likely to happen.*

- A thought representing a hypothetical scenario (e.g., the tap being open), gets miscategorized as a real memory—*I know the tap is closed, but thinking about it being open makes me feel worried and uncertain.*
- The same moral judgment is leveled at an *ego-dystonic* automatic thought (i.e., incompatible with the person's self-knowledge, goals, and needs), in the absence of an intention to act, as to an intentionally mendacious act—*thinking about cheating on my husband is as bad as doing it* (this is moral TAF, which also reflects inflated responsibility; see above).

The thinking error of *emotional reasoning* may contribute to an exaggerated importance being attached to thoughts: *Because the thoughts upset me, they must be important*, or *Because I'm anxious, there must be danger.* Here the emotions associated with a thought or the emotional context of a thought are mistakenly taken as sufficient evidence to support the validity of the thought.

Similarly, the cognitive-behavioral response to the thought can unreasonably (implicitly) be taken as evidence of its importance, such as where prolonged ritualistic rumination inflates the thought's significance: *It's important because I'm spending hours on it*, or where acting on the premise of the thought is taken to reinforce its truth value: *I avoid the situation because my fears are credible—and my fears are credible because I avoid it.*

Overimportance of thought control. This refers to exaggerated, unrealistic importance being attached to attaining complete control over one's thoughts (Purdon & Clark, 2002), such as by being able to immediately banish troubling thoughts from one's mind after having them, or avoid having them in the first place. The need for thought control tends to arise after attaching excessive importance to the thoughts and entails the use of control strategies such as thought suppression or mental ritualizing. The aim of these strategies is to avoid the negative consequences of inadequate control; as one client expressed, "If I can't stop myself thinking these [horrific] things, I'll lose my sanity. I think there is something ugly growing inside of me."

CLUSTER THREE

Perfectionism. Steketee (1999) defined perfectionism as "the tendency to believe that there is a perfect solution to problems, that doing something without mistakes is possible and desirable, and that even minor errors will have serious consequences" (p. 146).

Perfectionism can be expressed in beliefs such as *You either do the job 100% or not at all* and *To be able to respect myself, I can't afford any mistakes.* In OCD, perfectionism can manifest in varied ways. For example, one has to attain perfect certainty that adequate performance has been reached to avert feared consequences, such as being 100% certain that the windows are closed (*If they're left open, the house will be burgled*), or having a perfect understanding of the lecture (to be a good student), or knowing that every household item is perfectly in place (to feel in control). Perfectionistic standards can be applied to mental experience, actions, items, the self, or other people.

OCD perfectionism can manifest in the application of arbitrary or exaggerated standards for items, experiences, or behaviors, sometimes at the expense of or becoming a greater priority than their commonsensical function. For example, *My CD covers must not have any dust on them* may represent applying an exaggerated aesthetic standard that obscures or entails a failure to recognize that the CD covers are doing their primary job quite well—to label and protect the CDs. Or, in the example *When I click the light switch, it has to click in the right way by making the right sound*, the criterion of the sound of the click is arbitrary to the function of the click, which is to switch the light on and off (van Niekerk, 2009).

Intolerance of uncertainty. A state of uncertainty results when people with OCD conclude unreasonably (i.e., deviating from cultural norms or a flexible, functional viewpoint) that they do not have sufficient data to reach a conclusion of safety or allow closure. Eliminating or reducing the unpleasant state of uncertainty becomes the incentive for initiating an action, such as ritualizing, which is terminated once a sufficient level of certainty has been established.

Beliefs and appraisals in this category reflect the importance of establishing certainty, the view that this is necessary and attainable before one can proceed safely, along with fearful expectation of the likely negative consequences and risks inherent in not resolving uncertainty.

Other Beliefs

In addition to the above-mentioned beliefs, Wilhelm and Steketee (2006) propose considering two further beliefs relevant to OCD (and other disorders):

- Fear of positive experiences, such as predicting that positive experiences are risky or that improvement will not be sustained, which may be underpinned by a core belief: *I don't deserve anything good*. Please refer to the authors' detailed treatment manual for intervention guidelines.

- Fear of anxiety. Clients may worry that prolonged, intense anxiety would ensue if they do not perform their rituals. They may be concerned that ultimately this will lead to a loss of mental control, requiring them to be hospitalized. Alternatively, they may fear encroachment on their well-being (*It would spoil my holiday because I would keep on worrying that the safe is unlocked*) or on their ability to function adequately (*I'll be in a constant nervous state and will not concentrate properly*).

Cognitive Therapy for OCD

CT is a structured, present-problem-focused, short-term intervention that has been found to be effective in the treatment of a wide range of psychological disorders. It aims to assist clients in identifying, testing, and challenging unrealistic, unhelpful ways of thinking that underlie their problems and responding with accurate, helpful alternatives. It aims to modify counterproductive behavioral patterns that cause and maintain problems. Finally, it involves examining the underlying template of dysfunctional beliefs, which manifests in thought patterns across situations, and substituting it with a more balanced, fairer perspective.

All therapy work is administered in the context of a cognitive-behavioral case conceptualization, which is modified and updated as new data become available over the course of therapy. The case conceptualization, which is collaboratively developed and shared with the client, demonstrates the therapist's empathy for the client and offers the client a scientifically based alternative account of her problems. The client is an active participant, thus benefiting from being empowered to understand and eventually be able to apply effective therapy intervention.

Sessions are structured and usually sequenced in the following way:

1. Brief mood check and update on events since last meeting
2. Agenda setting and covering the main points
3. Setting home assignments (or deciding on an "action plan")
4. Summary and feedback

The course of therapy also tends to follow a set format, starting with an assessment during which the case conceptualization is developed, which guides the intervention. Initial work tends to focus on modification of situation-specific problematic thinking and behavior, with later work focused on addressing core beliefs and relapse prevention.

All the aforementioned features of CT also apply to its application in OCD. First, the client is thoroughly and comprehensively assessed. Hereafter follows client education and goal setting, distinguishing intrusive thoughts from their appraisals, cognitive restructuring and behavioral experimentation, and relapse prevention. Each will be considered in turn.

Assessment

Chapter 2 gave an outline of how to assess the client's obsessions and their triggers, compulsions and other safety strategies used, avoidant coping, and the feared consequences of facing the situation without using safety strategies. Building on and overlapping with the initial assessment, CT case conceptualization relies on further detailed assessment, as described below.

CASE CONCEPTUALIZATION

Collaborate with the client to decide which obsession to focus on first (if there are several). It may be helpful to focus on a more recent trigger situation (or, if there is a current high level of avoidance, a past or hypothetical encounter with a trigger), as this will tend to be fresher in the client's memory; this conceptualization can then be considered in light of how well it accounts for other occurrences of that obsession.

As previously described, the conceptualization guides intervention aimed at changing the appraisals that are instrumental in situational obsessive thinking, allowing maladaptive coping in these situations to be eliminated. Overlapping obsessions may benefit from the scaffolding of previous conceptualization—equivalent dysfunctional appraisal domains may be implicated; however, transitioning to working on a different obsession subtype may again require detailed assessment to allow tailored formulation and intervention.

The following table provides a summary of situational obsessive-compulsive features requiring assessment, to allow cognitive conceptualization (many of these data will already have been gathered during the initial assessment but may need to be updated, refined, and elaborated). These will inform the intervention and serve as outcome variables. For example, as therapy intervention progresses, you can assess on a 0 to 100 scale whether there has been a reduction in the belief in the appraisal that thoughts can cause catastrophic events to happen and a reduction in the urge to perform a neutralizing mental ritual.

Obsession features	Safety strategy features
External or internal triggers	Rituals (behavioral/mental) or other safety strategies used in response to obsessions (e.g., thought suppression, avoidance)
Content and frequency (e.g., number of occurrences per day/week) of obsessional thoughts/urges/images	Intensity of urge to engage in the ritual (0–100)
Emotions elicited (anxiety, but also guilt, shame, anger, irritation, disgust, or sadness—is there a dominant emotion?); (0–100)	Degree of success in resisting the ritual (0–100); which factors make it easier (e.g., feeling upbeat, or simply too depressed to care) or harder to resist (e.g., stress, tiredness, lack of social support)
Appraisals (or interpretation) of the obsession (see section below)	
Level of insight into the validity of obsessional fears and the need for the compulsion	

PSYCHOMETRIC ASSESSMENT

OCD symptom severity measures have already been considered in chapter 2. CT case conceptualization can benefit from administering the Obsessive Beliefs Questionnaire (OBQ; provided as an appendix in Frost & Steketee, 2002), a measure of the six appraisal domains considered above (see "Dysfunctional Appraisals in OCD").

Wilhelm and Steketee (2006) extended the original eighty-seven-item OBQ to include fifteen items assessing Consequences of Anxiety and fourteen items assessing Fear of Positive Experiences; the extended scale (OBQ-Ext) is provided in their book. The OCCWG (2005) has also developed an abbreviated OBQ (the OBQ-44), consisting of forty-four items that provide subscale scores for three appraisal domains (threat/responsibility, thoughts/thought control, perfectionism/uncertainty) and a total score. My preference is for the OBQ-44, although this lacks the added scope of the OBQ-Ext.

Administering the OBQ-44 or OBQ-Ext at various points in therapy—baseline, mid-therapy, and at the end—can be helpful in the following ways:

- Informing diagnostic questions
- Identifying appraisal/belief domains that are particularly relevant to the case conceptualization, therefore signposting focused assessment

- Allowing a quantitative assessment of treatment mechanism, thereby enabling tracking of treatment progress
- Providing quantitative evidence of treatment progress, or impediments to progress, which can be presented to the client, serving as encouragement or for troubleshooting

Despite these benefits, the heavy lifting in case conceptualization relies on careful questioning and data gathering during the assessment interviews and over the course of therapy and on the information from client self-monitoring. Psychometric assessment plays a smaller supplementary role.

Once an initial working model of the client's OCD has been developed (which will evolve over therapy) with preliminary identification of intrusive thoughts, obsession appraisals, and safety strategies in at least one trigger situation, it is possible to proceed to client education and goal setting.

Client Education and Goal Setting: The Cognitive Appraisal Model

First, briefly socialize the client into the generic CT model (see standard CBT manuals). The main idea is that the meanings we attach to an event contribute to our emotional and behavioral response. Inaccurate, counterproductive thinking leads to maladaptive behaviors and can be instrumental in the development of emotional disorder. Hereafter the focus shifts to educating the client about the cognitive appraisal model (CAM) for OCD, with a primary focus on its application to the client's obsessions.

When introducing the CAM, the style of the discussion is necessarily didactic, but as collaborative as possible. Ideally, this involves first building the model by asking questions of the client for obtaining information and eliciting feedback, then presenting a summary of the functional relationship between symptoms. Check in on the client's understanding of what has been presented, be on the lookout for ambivalence, clarify misunderstandings, and focus on building the therapeutic alliance.

The following template (inspired by Purdon, 2007) can serve as guidance (here applied to a client with a fear of contamination with toxic substances):

Therapist: In OCD, obsessions are taken to be the repetitive thoughts that cause you to experience anxiety or distress. These feelings make sense given what the thoughts represent to you. The obsession signals that you could be in danger—that you could have touched an item that contains lead or other poisonous chemicals and that your health could be damaged, or that you could poison someone else.

And that's why your responses make sense—the obsession signals a problem and you then have to deal with it, right? And this is what you have described: first, you're on the lookout for items that could contain poisonous substances and you avoid handling them—let's call this avoidant coping. And when you have to touch an item you're unsure of, you wear gloves and then get rid of them afterward—more avoidance. You also wash your hands frequently, to the point where the skin is red and flaky, and have frequent long showers. And you do hours of research on items you purchased, to investigate their toxicity. These are your rituals or compulsions. Finally, in desperation, sometimes you try to clear your mind of the obsessions by trying to distract yourself with positive thoughts.

Now these strategies help in the short run; you feel safer and less anxious, or at least less anxious than how you would have expected to feel if you did nothing. But they exact a toll—costing you time and effort, causing hassle, sometimes making you feel miserable—and that's why you're seeking help, right?

But not only do these coping strategies cost you, they also make your problem worse. This is how it works: First, when you avoid triggers of your obsessions or perform compulsions, you do not experience the obsession in its distressing form. Therefore, you don't allow yourself the chance to learn how dangerous having the obsessive thought actually is. Maybe having the obsession is completely safe and you don't need to do anything at all, but at the moment there is no way for you to know if this is the case. In fact, it may even seem that when you do the compulsions, nothing bad happened because you did them, or that the risks would be much higher if you didn't use the strategies. Further, when you avoid obsession triggers, it may seem as if bad things didn't happen because you didn't have the obsession, or as if the situation was safe because you didn't have a thought about danger.

Second, the mind has a habit of tying up closely related experiences into a single memory, resulting in one thing reminding you of the other, even if you don't want it. Because the compulsion is so closely linked with the obsession, it will eventually inadvertently trigger reminders of the obsession, resulting in having more obsessive thoughts, which then cause you distress and require you to act, and so on. Therefore, handwashing, showering, and brushing off your clothing, and anything linked to the compulsions, such as soap and hand towels, become likely triggers for the obsessions. In a similar upside-down way, attempts to avoid obsession triggers can result in triggering the obsession. Furthermore, when you try to clear your mind of obsessions, what you try to substitute them with—such as positive thoughts—gets linked with the obsession and starts serving as a reminder of the obsession, leading to having more obsessions, rather than less.

Third, thought suppression tends to be a raw deal for another reason: to not have certain thoughts, the mind has to keep an instruction up and running about which thoughts to look out for, in order not to have them. But this instruction then paradoxically triggers the obsessions. Research shows that nobody can entirely suppress unwanted thoughts. And when you don't succeed in controlling the thoughts, this might feed further into your anxiety. Therefore, overall, not trying to control thoughts tends to be more helpful.

A key element in the whole cycle is how you interpret your obsessional thoughts, or which meaning you attach to them. If you didn't consider them important or meaningful, you wouldn't feel anxious and you wouldn't need to use any disruptive coping strategies. That would put an end to the OCD.

What we know from research is that most people without OCD get a wide range of intrusive thoughts, some odd-seeming and even violent, but they don't develop into OCD obsessions [*show the client a list*]. What distinguishes people with OCD is that their interpretation of the thoughts essentially results in the thought being "weaponized."

To give you a sense of what I mean, let's select a non-OCD intrusive thought you have experienced [*have the client select a thought*]. Okay, so you're at the station and you're thinking, I'm going to push that woman in front of me onto the tracks. How would you interpret this thought to cause you alarm? Your job here is to weaponize the thought.

Client: This is my violent subconscious coming out!

Therapist: How does this interpretation make you feel and how would you act?

Client: Definitely worried. I'll step right back, maybe leave the station.

Therapist: Okay, so here the interpretation had an important influence on your emotion and your actions. Now how would you have interpreted the thought as something that can safely be ignored?

Client: Well, if I don't want to push the woman, then I won't do it, so I can ignore it?

Therapist: Yes, and this illustration points to one key characteristic of people with OCD—they interpret their thoughts as being powerful, highly meaningful and potentially harmful, unlike those who don't have OCD, who interpret their thoughts as just that—harmless, insubstantial thoughts.

So how do we treat OCD? Given the central importance of the interpretation (or appraisal) of your obsessive thoughts, we need to slow it down so that we can examine the embedded assumptions. The interpretations in OCD are incorrect and extreme. Because eventually they happen almost automatically—the mind takes shortcuts—they have eluded careful examination. So, in treatment, we will identify the automatic interpretations of your obsessions that have led you to believe that something is threatening or dangerous. We will carefully reflect on whether they are accurate. You will modify the inaccurate ones and replace these with a balanced and realistic interpretation of your thoughts. Finally, we can reflect on the new balanced appraisals and reconsider whether your current OCD coping strategies are in fact necessary at all.

The discussion can then bridge to collaborative consideration of treatment goals. By understanding the conceptualization, the client will be able to make suggestions about appropriate goals for symptom reduction and the mechanisms that maintain symptoms.

Hereafter the focus shifts to assisting the client in developing competency to identify obsession appraisals. This sometimes tricky issue is examined in more depth next.

Distinguishing Intrusive Thoughts from Their Appraisals

Many of my trainees (and colleagues) over the years have scratched their heads trying to distinguish the initial intrusive thought from its appraisal, and in some presentations, this distinction is not intuitive or clear-cut, let alone for a client who may have been immersed in her complex OCD for years. Nevertheless, the distinction is important: in the CAM, "appraisal" has been identified as, and became code for, those cognitions which can usefully be addressed using existing CT tools. The therapy guideline is to address the appraisal of the obsessive thought and not get entangled in unproductive discussions about whether the obsession is true. That is, focus on the meanings assigned to the thought, not on the thought content.

However, the clinical reality is sometimes more complex. Wilhelm and Steketee (2006) suggest that there are instances where working on the obsessional content directly can be useful (such as when addressing overestimation of danger).

It is also useful to bear in mind that in the CAM literature, the "appraisal of the *thought*" can refer to reaching a conclusion based on the fact of having had a certain thought (or metacognitive inference; see chapter 6): Client A: "I have upsetting, obscene thoughts when in church (intrusive thoughts), which *must mean I'm a bad person* (appraisal of the thoughts)."

However, somewhat counterintuitively, the appraisal of the thought can also refer to a calculation and prediction of the possible feared consequences over time if the conclusion represented in the initial thought (which already represents the appraisal *of a situation* as posing a threat) were true. Client B: "I touched the paper bin and worried that I got germs on my hands (danger appraisal 1: *There are germs on the bin*; danger appraisal 2: *Now they could be on my hands*) and that I could infect someone else (danger appraisal 3) and keep blaming myself (danger appraisal 4). That's why I have to act (responsibility appraisal 5)." In this example, the CT focus will tend to be downstream in the chain of reasoning, on appraisals 3 to 5.

When queried about the original locus of danger, Client A would likely identify his thoughts, whereas Client B might point to the bin rather than to her thoughts. In the case of Client B, this would go against a view of the problem being completely, satisfactorily described as being the appraisal of thoughts because the subsequent appraisal of the situation seems to be what is at stake. (If this convoluted conceptual issue is up your alley, it may be worth considering the distinction proposed by Lee and Kwon [2003] between autogenous and reactive obsessions.)

It is also worth remembering that how the client experiences and describes his obsessions does not necessarily map readily onto a simple scheme of first having an intrusion, which is then followed by an appraisal and use of a safety strategy. For example, one client, suffering from obsessions about having made mistakes that caused harm to others, described having crossover between triggers and appraisals (i.e., the appraisal can appear to be the first thought, and snippets of ritualistic self-reassuring can serve as obsession triggers). Another example is someone who denies having intrusive thoughts but engages in extensive ritualizing. Here, the therapist's role is to guide the client in distinguishing the upstream and downstream components according to the case conceptualization and focus intervention on the appraisals that foster maladaptive coping. Therefore, it is important to give effective guidance to clients to allow them to become adept at identifying their dysfunctional appraisals before proceeding with intervention to change them.

QUESTIONS TO IDENTIFY APPRAISALS

Purdon and Clark (2005) and Clark and Beck (2010) offer helpful suggestions for questions that allow obsession appraisals to be identified:

- What makes the obsession hard to ignore, or important?
- What does it mean about your personality or moral character?
- What does it mean for the near (or distant) future?
- What will happen if you don't ritualize, avoid, or suppress (i.e., perform safety/control strategies)? What is the worst that could happen if you didn't do [the safety strategies]? (This question is particularly helpful in the absence of an intrusive thought trigger for the compulsion.)
- Why does the obsession keep coming back?
- What will happen if you can't get rid of the thought?

For example, one client worried that he might impulsively have contaminated a saucepan with a toxic agent in the period since last use (intrusion); his appraisal was that because he thought what he did, and felt anxious and uncertain, he could actually have done it and therefore could poison himself if he did not rinse the saucepan. Another client described a worry that she had not tied her shoelace in the right way (intrusion); her appraisal was that she would have bad luck, play tennis badly, and lose the match if she did not re-tie her shoelace in the "correct" way.

When identifying appraisals in the context of an obsession that apparently elicits several emotions, careful exploration can reveal which thoughts are "behind each emotion." For example, if the client says that she felt both anxious and angry after an obsession "spiked," representing fears of being a miscreant, questioning may reveal that her anxiety is linked to an appraisal of the fears being valid with a consequence of ending up rejected and isolated. Questioning might also uncover that her anger is linked to a train of thought about the fears "being OCD" and how unfair it would be for others to stigmatize her.

Following questioning and exploring the meaning of the obsession with the client, draw up a list of reasons why the obsession poses a significant personal threat and review these with the client.

THOUGHT RECORDS

The client can also learn to identify the appraisal from the completion of thought records between sessions. Initially a simple, four-column form with columns for situation/intrusion trigger, intrusive thoughts, emotion, and interpretation will suffice (or use the initial columns of the Intrusive Thoughts Worksheet in appendix 1, also available at http://www.newharbinger.com/38952), which ideally should be completed daily, when obsessions are triggered. These can be reviewed at sessions.

When the client can accurately identify the appraisal, this enables transition to the next phase: cognitive restructuring of the appraisal.

Cognitive Restructuring

All the cognitive-behavioral strategies used for cognitive restructuring in standard CT have an application in treatment for OCD. Inform the client that the aim of therapy is to examine all the evidence and reach a balanced conclusion in terms of whether it supports one of two basic theories:

> *Theory A*: The obsession is accurate, with cause for concern and action (embodying the unhelpful appraisal: there is a *danger problem*).
>
> *Theory B*: The client is someone who is anxious that the obsession may be accurate and therefore responds in unhelpful, counterproductive ways (the helpful appraisal: a *worry problem*); in this case, the appropriate response to the obsession is to do nothing.

Take note that toward the later stages of therapy, when encountering a trigger, the process of examining the evidence has to transition to making a rapid decision, similar to non-OCD decision making. This is to prevent evidence review from becoming ritualized, which happens when covering ground that has been covered before. Background knowledge does not require repeated, detailed reexamination. (When leaving home in the morning, one doesn't need to remind oneself that crossing a road is dangerous and which safety measures to employ.)

Next, I will consider some of the key general strategies used, followed by their application to specific appraisal categories.

GENERAL STRATEGIES

Socratic questioning. This key technique entails the therapist asking a series of questions with the aim of encouraging clients to examine the validity and utility of their thoughts and conclusions. The rationale for using a more interactive, questioning approach, rather than a more directive didactic approach, is that it reduces the risk of a power struggle and the potential activation of what might be a client's overgeneralized, negative interpersonal beliefs (e.g., where the therapist is appraised as yet another critical, invalidating figure). The assumption is that knowledge derived from self-reflection will be more convincing. Examples of appropriate questions are:

What is the evidence for and against?

Is there a different way of looking at this?

What are the pros and cons of this attitude?

What would be the consequence if you changed this belief?

Identifying unhelpful thinking styles (inspired by Beck, 1976, 1995; also see Ellis & Ellis, 2011). The table below provides a description of common self-sabotaging thinking styles, each with an example and consideration of the flaws in this thinking, to help lead the client to a more balanced viewpoint.

Unhelpful thinking styles	Examples	The drawbacks in this thinking
All-or-nothing thinking: the tendency to evaluate one's own or other people's performance or personal qualities as black or white. Also called the *fallacy of bifurcation* (Pirie, 2006).	*My life is either under control or in complete chaos.*	The world is *complex*: people or events are usually not just one way or the other. "Shades of gray" are not acknowledged.
Magnification: a tendency to reach blanket negative conclusions about oneself or other people on the basis of flimsy evidence.	After making a single mistake: *I'm useless.*	This is not a balanced conclusion reached by looking at *all the evidence* objectively.

Minimization: a tendency to discount any information that doesn't fit with one's negative view of oneself or other people.	After successfully resisting ritualizing: *I could only cope because the therapist was there.*	The same as for magnification.
Overgeneralization: when one arbitrarily concludes that a negative event will happen over and over again.	After your child left the window open: *You can't trust anybody.*	A conclusion is reached that a negative pattern will continue, but looking at all the relevant evidence suggests otherwise.
Catastrophizing: dwelling on the negatives in a situation and believing things are worse than they actually are.	*My OCD has ruined my life.*	There is insufficient attention to aspects of the situation that contradict one's negative conclusion (there may well be parts of one's life that are relatively unaffected by OCD).
Low frustration tolerance: the belief that something is so bad that it simply can't be tolerated; this is frequently linked to catastrophizing.	*Being anxious is simply dreadful and must be avoided at all costs.*	On closer scrutiny, the belief goes against the evidence that the body and mind are strong. It's better to decatastrophize, face the situation, and accept the inescapable discomforts that are part of the fabric of life.
Inflexible demands: making rigid demands about things out of one's control (and then catastrophizing when the demand is not satisfied).	*I simply couldn't look myself in the mirror if I kept on having thoughts of attraction to women other than my wife.*	Unnecessary upset is caused by the failure to recognize that not everything in life is 100% under one's control. Being upset does not help you cope with the situation in the best way.

Unhelpful thinking styles	Examples	The drawbacks in this thinking
Emotional reasoning: when one's emotions in or after a situation dictate the conclusion about oneself or the situation.	*I'm just feeling so guilty—I must have done something really bad.*	The emotion resulted from a cause other than a realistic view of the situation; this leads to a biased conclusion about the situation.
Negative prediction (or "crystal ball gazing"): acceptance of future predictions as fact.	*Treatment won't work, so what's the point of trying?*	The future is never 100% certain. Rather than make negative predictions, take action to help positive predictions come true.
Personalization: arbitrarily concluding that an event is personally relevant, or "taking things too personally."	After a friend did not respond to your text message: *She has rejected me because I have OCD.*	Other possible explanations for an event are unreasonably excluded (e.g., maybe she hasn't had time to respond).
Mind reading: assuming that one knows what others think.	*They don't want to see me because they think I'm a pervert.*	It is more helpful to not go beyond the available evidence; instead, accept a lack of certainty or verify by asking.

Downward arrow. This technique is aimed at uncovering meanings and self- or other-beliefs underlying the client's thinking about events. Although the aim is primarily one of assessment and formulation by identifying the underlying cognitions, it can also engender belief change through inviting the client to reflect on his *articulated* assumptions (which, by definition, are more at the fore and more visible than implicit assumptions). The technique requires asking a series of probing questions, such as:

- What would/did that mean to you?
- What was/would be bad about that?
- What would/did that mean *about you*? (This modified phrasing is aimed specifically at identifying self-beliefs.)

Initially the therapist would ask the questions, but ultimately the aim is to educate clients to do the same for themselves.

Therapist: So, what is bad about wanting to look at the child?

Client: Maybe I enjoyed it (*emotions: guilt, anxiety*).

Therapist: What would that mean?

Client: There could have been some sexual attraction.

Therapist: What would that mean about you?

Client: That I'm a creep, you know, a child molester.

Therapist: And in your mind, what would be so bad about that?

Client: I'd be a danger to my own children. I might lose them.

Questioning can be terminated when a satisfactory account of the client's emotional experience in the situation has been achieved and when further questioning did not yield additional evidence.

Continuum technique. This is particularly helpful where the client has a tendency to make pejorative judgments that express dichotomous categories about the self or others. Consider a client who has an obsessional fear about her daughter accidentally swallowing medication.

Therapist: What would it mean about you if it happened?

Client: I'm a bad mother.

With this client, one use of the continuum technique would involve collaboratively developing clear definitions of which traits and behavior would describe a 100% bad versus 100% good mother, and having her place herself on a line connecting the two labels (a bidirectional continuum). Clients usually assign a rating that is a distance away from the polar worst outcome (and if not, the client's appraisal of the evidence can be further explored).

However, Padesky (1994) pointed to possible advantages for developing the adaptive belief by using a unidirectional continuum—in the example, this would be a line measuring being a good mother, from 0 to 100%. On this continuum, any positive shift would represent having *more of a desirable quality*, which may resonate more than a positive shift on the bidirectional continuum (e.g., from a 100% bad to a 50% bad mother), which may only signify having *less of a negative quality*. Whichever strategy is used, the point is to prompt the use of evaluative criteria that capture reality fairly by giving an account of the full breadth of data.

Hypothetical roles and interactions. This strategy is aimed at presenting clients with a helpful alternative vantage point on their behaviors and beliefs. For example, you might say, "If your friend blamed herself harshly after having made [the mistake that the client fears making], what would you say to her?" The client's response may allow you to raise the possibility of *double standards.* Or, it can be powerful to ask clients what they would advise their own children, or a younger person they mentored.

A courtroom scenario can be envisaged where the client becomes the prosecuting attorney (representing the OCD) implicating herself, and the therapist is the defense attorney. Roles can be reversed at some point during the exercise, the strength of the arguments can be compared, and a prediction can be made of who the jury might agree with. Or the client can be invited to be the judge responding to the therapist as prosecuting attorney (which introduces the useful concept of "hard evidence" and legal criteria for admissible evidence). Ultimately, the aim is to introduce new concepts to benefit the client's analysis and encourage a more flexible vantage point.

Behavioral experiments. This powerful and important strategy for cognitive restructuring involves intentionally modifying problem behavior to test whether a client's prediction about the likely negative consequence is borne out. It is particularly appropriate when testable hypotheses can be derived from the client's beliefs, such as when the client makes a conditional assumption where the time frame of the prediction lends itself to verification ("If I don't check each of my windows three times, my anxiety will be unbearable all day and I wouldn't be able to cope at work").

A behavioral experiment might proceed according to the following steps:

Step 1. The therapist and client agree to test the client's belief about what would happen if, when leaving for work in the morning, he closed an open window but did not check windows that had previously not been opened.

Therapist: Let's find out how well your thinking maps onto reality. You have been acting as if there is a very high likelihood that your anxiety will be unbearable and remain very high throughout the day if you didn't check, and that you wouldn't be able to cope at work. Of course, this scenario is possible, but let's find out what happens in reality, to help you let your actions be guided by accurate predictions.

Step 2. The client rates the probability of two predictions: (a) that his anxiety will remain steadily above 80 on a 0 to 100 scale over the course of his working day; and (b) that he will not be able to cope with a list of routine tasks over the day. The client records these predictions on a Behavioral Experiment Worksheet (appendix 1, also available for download at http://www.newharbinger.com/38952), which also states unambiguously which evidence will support the feared outcome and how this will be measured (e.g., the client can use a cell phone application to prompt him to rate his anxiety every hour).

Step 3. The client executes the behavioral experiment (i.e., not checking windows in the morning that had not been opened), records the outcome (as applies to the predictions outlined in step 2), and reflects on the meaning of the new data. Clients can benefit from reflecting on the implication of the new data produced by the experiment for (a) their thoughts about the situation or that category of situation ("Do the findings of the experiment support a theory of this being a worry problem or a danger problem?" "Do you think there may be an application to other similar situations?") and (b) their OCD ("What have you learned about the OCD's con job?"). The emphasis is on *learning*, not outcomes: "Let's say you found that you had a complete anxious breakdown at work and couldn't complete a single task; that would definitely be upsetting, but how do you think you could benefit from this knowledge?" (The empirical ethos is emphasized!)

Between-session therapy work. This is a standard feature of CT work, and the aim is to extend and consolidate treatment work between sessions. This encourages client autonomy and steepens the learning curve. However, despite the therapeutic potential of between-session work, it is frequently underutilized. I, for one, can testify to numerous instances of having been either overambitious or underambitious in suggesting between-session work, or throwing in the towel too soon when clients do not engage fully with home assignments. Therefore, it is worth heeding Padesky and Greenberger's (1995) guidelines:

- Make sure assignments do not exceed the client's skills or capabilities at that point in therapy.
- Create assignments that are collaboratively developed, relevant, and interesting (not as easy as it may seem!).
- Provide a written summary and remember to follow up on the assignment at the next appointment. The emphasis is on learning rather than outcome attainment (e.g., learning which strategies are helpful for overcoming low motivation to exercise, rather than exercising per se).
- Empathetically explore reasons for noncollaboration, and problem solve as appropriate. Keep in mind that clients are frequently ambivalent about homework and may not have stated this openly at the time (see the section on motivational interviewing techniques in chapter 3).

Next I will consider cognitive restructuring in specific categories of appraisal. This work can be supplemented by the client's use of the Intrusive Thoughts Worksheet (appendix 1, also available at http//:www.newharbinger.com/38952), with columns for situation/intrusion-trigger, intrusive thought, OCD interpretation, emotion, OCD coping strategies, and helpful interpretation and response. However, the worksheet should only be an initial therapy prop to assist the client in formulating the situation and should eventually be withdrawn in favor of an approach that resembles non-OCD thinking and action: nonexcessive, flexible, and adaptive—predicated on the demands of reality and sound judgment.

OVERESTIMATION OF THREAT

Inflated estimates of the probability and magnitude of danger are common in OCD, particularly in contamination or checking subtypes. The strategies discussed below may prove to be useful, although obsessional symptoms may prevail if additional appraisals of inflated responsibility or intolerance of uncertainty are unattended to ("Even though I appreciate the risk is one in a billion, I still can't stand the thought of potentially being to blame for someone getting hurt!").

Cognitive strategies. Consider the following lines of questioning and data-gathering strategies:

- How would a reasonable, responsible, trusted friend or relative (ask for the person's name to make the question more personal) estimate the likelihood of the negative event? (The client could also conduct a survey, asking specific, detailed questions of trusted others, and collate the information.) Which factors would explain the discrepancy between that person's estimate and the client's estimate?

- The client may benefit from consulting experts, such as an infectious disease physician to discuss the probability of contracting HIV from using a public toilet, or an occupational health physician to discuss concerns about asbestos exposure. It may be advisable to have a conversation with the expert to brief him or her about OCD prior to the meeting with the client.

  This strategy is likely to be particularly helpful when the client is ignorant of relevant information and considers the expert to be credible; however, this intervention can easily be sabotaged by obsessional doubting ("I didn't provide all the details, which might have changed his opinion"). Therefore, it would be wise to prompt the client to preconsider what level of detail is reasonable and how the new information can help her modify her appraisals and behaviors to overcome OCD (inflated responsibility and intolerance of uncertainty are also relevant to consider here; see below).

- What would one expect the world to look like if the client's estimate were accurate? If there were a good chance that red spots in public places were in fact infected blood posing a serious risk to health, might all cleaners be required to wear protective suits and have to pay sky-high life insurance premiums? If actuaries made the wrong call about calculating risk, this would cost insurance companies a lot of money!
- Is the client ignoring risk-mitigating factors, such as the immune system preventing disease, or built-in mechanical safeguards such as an auto switch turning off the coffee machine?
- Is there a possibility the client may hold a negative prejudice against himself, expecting bad things being particularly likely to happen to him? What alternative perspective would represent a measured assessment of *all* the evidence? (See "Core Beliefs" below.)
- Past behavior is a reasonable basis for predicting future behavior—what does the client's history of coping with adversity suggest about her ability to cope with future adversity?
- Look out for unhelpful thinking styles evident in the client's reasoning about the extent and consequence of future adversity, for himself and others, such as inflexible demands ("This simply can't be allowed to happen"), catastrophizing ("That would ruin my life"), dichotomous thinking ("I will be a failure"), and low frustration tolerance ("I wouldn't survive it!"). (Some good YouTube clips are available in which the mercurial CBT pioneer, Albert Ellis, discusses thinking errors.)
- It may be helpful to introduce the client to the correct statistical method for calculating probabilities (van Oppen & Arntz, 1994):
  1. Obtain a percentage estimate from the client for the feared consequence.
  2. Identify all the steps in the causal chain required for a feared event to occur.
  3. Estimate the probability for each step, or condition, to occur (accessing relevant published statistics can make a useful contribution here, such as local incidence or prevalence rates for a disease, house fires, or burglary).

4. Multiply the probabilities for each step to derive an estimate of the feared outcome.
5. Have the client reflect on the discrepancy between her initial estimate and the calculated estimate.

For example, Beth's concern was that the office building would burn down if she left her computer switched on, which was the rationale for her checking it repeatedly. She and her therapist identified the following conditions that would have to apply for the dreaded event to occur and their probabilities: (1) the computer's electrical system being faulty: 1/1,000 chance; (2) the sprinkler system malfunctioning: 1/100; (3) the night watchman not being on his post: 1/100; (4) passersby noticing the fire too late: 1/3. Therefore, the calculated odds for the building burning down were 1/1,000 x 1/100 x 1/100 x 1/3 = 1/30,000,000!

This method can be particularly helpful when the client holds a misconception about probability calculation, but it can activate intolerance of uncertainty concerns (e.g., about which probabilities apply for individual links in the chain), and therefore its use can develop into a cumbersome neutralizing strategy, which will have to be addressed. Ultimately, the aim is to develop a quick, commonsensical understanding of how probabilities are correctly calculated to better appreciate the equations that underpin acceptable risk.

Behavioral experiments. Overestimation of threat frequently lends itself well to behavioral experimentation. This may overlap with ERP (see chapter 3), but in CT the aim is belief change rather than deconditioning. For example, the client can write phrases rather than full sentences in a text message to a colleague to see if this is catastrophically misunderstood, or not take her receipt at the store auto checkout machine and see if she gets arrested. The steps for behavioral experimentation, outlined above, apply each time.

INFLATED RESPONSIBILITY

Markers of inflated responsibility appraisals are when clients' thinking reflects elements of self-blame and self-recrimination, with associated feelings of guilt and shame, owing to perceiving themselves as not having acted morally to prevent harm. Responsibility and self-blaming interpretations can reasonably derive from overestimation of threat. If the latter has been successfully addressed, the former does not require intervention. However, *inflated* responsibility may require additional intervention, such as when the person's thinking is skewed toward *excessive* responsibility taking and self-reproach—particularly if inflated responsibility beliefs manifest in symptoms that prevail after overestimation of threat has been corrected.

Therefore, dysfunctional thinking can manifest at four levels in the reasoning chain: (1) threat estimation (see previous section), (2) assumption of *causal responsibility* for a negative outcome (i.e., believing that one's action contributed to the negative outcome), (3) assumption of *moral responsibility* (leveling moral judgment on one's behavior), and (4) estimation of negative consequences (e.g., personal or social). Dysfunctional thinking would be represented at multiple levels by overestimation of threat (e.g., "There is a high likelihood a child will step on or play with that piece of rusty wire lying on the road, scratch himself, develop blood poisoning, and die"), erroneous or inflated causal responsibility ("If I don't pick it up, I will cause someone's death"), inflated negative moral evaluation ("That would make me a murderer"), and inflated negative consequences ("I'd have to kill myself because I couldn't live with the guilt"). Alternatively, unhelpful thinking may manifest selectively at one or two levels only, but the client's distress will likely escalate when more levels are implicated. Assumption of causal responsibility would usually be a precondition for assuming moral responsibility and anticipating negative consequences.

Cognitive and behavioral strategies. Cognitive restructuring techniques can address dysfunctional thinking at all four levels listed above. Strategies for addressing overestimation of threat have already been considered.

The following may be useful in addressing appraisals of erroneous or inflated causal responsibility (and by extension, subsequent appraisals, including moral responsibility):

- When there is a likelihood of thought-action fusion (TAF), consider the evidence for the existence of "mechanism" (e.g., "How" can a thought make a plane crash?) and use behavioral experiments. For example, have the client try to make your pen run dry by thinking about it, then ask, "What does the outcome suggest about the power of thoughts?" (See "Overimportance of Thoughts" below for more detail.)

- Being the sole person responsible is not the same as sharing the responsibility with multiple others. The pie chart technique (van Oppen & Arntz, 1994) can be helpful here:

  1. Draw a circle to represent the causal responsibility for the event, and ask the client to estimate her percentage of the responsibility (e.g., the client rated her responsibility for causing a fire in her office building due to leaving her computer switched on as 100%).

  2. Ask the client to inclusively think of all the possible people or other entities who influenced the event and list them, including herself. (It

is worth considering a contribution of cultural factors to this reasoning; for example, clients from more collectivist societies such as East Asia may conceptualize responsibility more inclusively.)

3. Have the client assign a percentage to each person or entity to represent their degree of influence and then rerate her own (the computer company responsible for faulty wiring: 40%; the servicing company responsible for the sprinkler system not working: 30%; the caretaker not being on his post: 15%; client: 15%; see figure 4.2).

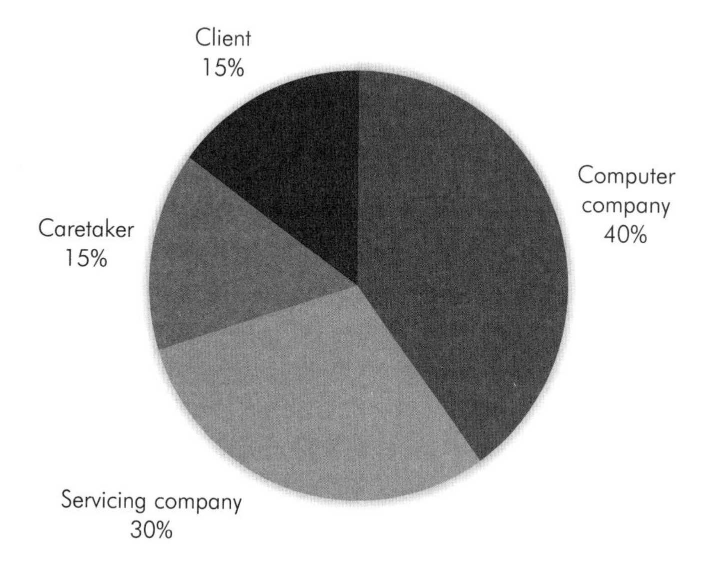

Figure 4.2. Pie chart technique

4. Reflect with the client on the discrepancy between the initial and final estimates of personal responsibility; the aims are to reach a fairer conclusion by acknowledging the contribution of others to causal or moral responsibility for a negative outcome—*any* responsibility is not the same as *all* responsibility.

The following may be useful to address overblown moral self-criticism:

- Ask what the client would tell his best friend (or child) if he described self-reproach about having been (causally) responsible for the negative outcome. Would the client be able to offer a sympathetic view? If so, what does the client make of his own double standards?
- Prompt the client to reflect on blame-mitigating factors when these are being ignored, such as in the situation described above, where a client

worried about leaving a piece of rusty wire on the road, resulting in a child suffering harm after stepping onto it. Here there may be a lack of consideration of the following blame-mitigating factors:

- The client did not intend to cause harm.
- She was not assigned the responsibility to clean the road or look after children.
- There were no clear signs ("hard evidence") of danger in the situation (e.g., noticing barefoot children playing on the road, or a young child picking up the piece of wire).
- There are other parties to bear the burden of blame if a child were harmed (e.g., the person who dropped the wire, the caregiver, parents; see pie chart technique above).

- Have the client contemplate what society expects of the behavior of a reasonable person when faced with a similar situation, similar designated responsibilities, and a similar level of knowledge at the time of acting (the client can conduct a survey of trusted others to explore this question). Any situation presents a thousand opportunities for taking action to prevent remote possibilities of harm to the self or others, which is why society adopts a compromise by recommending action when there is *real and evident danger or risk*. Further, what is the basis for acting only on *this* possibility but not the others (e.g., remove the piece of wire but not the sharp stones)?
- To address dichotomous thinking, present the client with a culpability continuum illustrating the cause of harm:

An unintentional, unforeseeable honest mistake	————————————	Premeditated, brutal murder of an innocent victim

Where would her action fall on the line?

Where appropriate, the fourth level of dysfunctional thinking, inflated negative consequences, can be addressed using Socratic questioning, by considering the evidence for and against the client's inaccurate belief and formulating an alternative, more realistic and more proportionate perspective.

Behavioral experiments. A "transfer of responsibility" experiment can be useful in demonstrating the influence of responsibility appraisals on the client's symptoms. The therapist "assumes responsibility," in writing, for any harm that may occur if the client did not perform rituals for a defined period, during which the client assesses the impact on his symptoms (Rachman, 2003b). Many clients may find it easier than bearing the full burden of perceived responsibility and experience a reduction in their symptoms during this period, demonstrating the significant role of this category of appraisal.

OVERIMPORTANCE OF THOUGHTS

Maladaptive appraisals in this category occur when individuals with OCD misinterpret characteristics of the obsession, such as the psychological context (e.g., not appreciating the absence of an intention to act on the thought, a desire to have the thought, reward entailed in the behavior represented in the thought, memory for the action represented in the thought, or a realistic, credible rationale supporting the thought), emotional consequence (e.g., thoughts seeming more believable and acquiring emotional resonance through causing distress), or form (e.g., thoughts seeming more believable because of repetition). These misapprehensions become the basis for the client's inaccurate predictions or conclusions about

- their past or future behavior (e.g., "I suddenly worried I did something offensive" or "I could violently lash out"; these reflect probability TAF),
- their moral character or personality ("I keep on thinking that I'm married to Jesus, which is blaspheming"), or
- catastrophic remote events ("I thought about my nephew being in a car crash, which might put him in danger through bad luck"; probability TAF).

(The idea of fusions is further developed in MCT in chapter 6).

These conclusions deviate from a scientific viewpoint and what would be culturally appropriate. For example, research shows that nonclinical intrusive thoughts that deviate from the person's intentions and values are common and not a basis for concern. Furthermore, appraisals instrumental in religious obsessions (e.g., that an unwanted, intrusive thought is as sinful as an intentional, desired action) tend to digress from or exaggerate what an authority in the client's religious community would subscribe to. Or, these may reflect an OCD-contrived and -facilitating metaphysical system conceived in the absence of religious affiliation.

Cognitive strategies. The following may be helpful to address overimportance of thoughts:

Psychoeducation. The aim is the provision of accurate knowledge about psychological functioning and normalizing of distressing intrusive thoughts, which can be accomplished in the following ways:

- Ask the client which intrusive thoughts she expects to be experienced in the population at large. Then compare this with a handout listing intrusive thoughts that have been reported by typical individuals (see appendix 2; also available at http://www.newharbinger.com/38952) or revisit similar insights presented to the client previously.
- Suggest that the client conduct a survey of trusted others, querying them about intrusive, "immoral" thoughts they may have experienced.
- If there is evidence of sexual obsessions and misconceptions about sexual behavior, the client may benefit from reading authoritative texts such as *The Mind and Heart in Human Sexual Behavior* by Alan Bell (1997).
- Client fears about going mad may be based on an erroneous understanding of relevant diagnoses; therefore, clarify diagnoses as needed.

Socratic questioning. If the client is concerned about the obsession translating into unwanted action, prompt him to consider which conditions need to apply for this to happen, for example:

1. a thought with content signifying an action
2. a conclusion that the action is desirable (i.e., that the pros of acting outweigh the cons)
3. a conclusion that the action is achievable
4. an *intention* to act on the thought

You could put it to the client that a thought about an action, in the absence of an intention to act, is like a car without wheels—should one be worried about crashing it? The question of whether the client has an intention to act, or not, may elicit obsessional doubting ("I know I don't, but..."). Therefore, it may be useful to reflect on what is a reasonable basis for verifying the presence of *intentions* about conclusive, extreme action, as may be represented in aggressive or sexual obsessions. The reality is that the client immediately knows whether an intention exists or not (similar to non-OCD extreme action).

It may also be useful to consider what factors recently inhibited a client's acting on thoughts about *desirable* behavior. Should those factors be considered to be more potent or less potent in the case of inhibition of grotesquely *undesirable* action? Or if the client's concerns are based on past impulsive behaviors, the discussion can reflect on the difference between the unwanted behavior represented in the obsession and the reward incentives that motivate impulsive or habitual behaviors.

This has already been referred to under "Inflated Responsibility," but it could be useful to query the mechanism whereby thoughts can influence external events (e.g., how thinking of someone getting ill can make him ill). You can explain the concept of mechanism to the client using an analogy, such as why does ice cream melt when taken out of the freezer? The therapist can also explore with the client the merits of scientific versus nonscientific explanations, which can be illustrated with examples from a domain of the client's life unaffected by OCD (discussed further in "Intolerance of Uncertainty" below).

A strategy that is helpful for moral TAF is presenting the client with a line drawn on paper, placing the most immoral person the client can think of at one end of the line and the most moral person at the other end (from Purdon & Clark, 2005). First, ask her to place herself and a good friend on the line. Then, ask where she would place her friend (1) if she experienced similar obsessive thoughts to the client, or (2) if she had the thoughts *and* acted on them. This visual metaphor explores the distinction between the moral implication of thoughts only, versus thoughts + intentions, versus thoughts + intentions + behavior. Which should weigh more in terms of moral admonishment?

Behavioral experiments. Behavioral experimentation can be a useful intervention for overimportance of thoughts, particularly in these situations:

- Likelihood TAF can powerfully be addressed within a behavioral experiment, where this is practical. For example, the client can experiment with walking past a vulnerable person while thinking thoughts about acting offensively and assess what actually happens (the experiments can also follow a graded format, with the client initially walking past the therapist, thinking of acting offensively, then progressing to a more anxiety-provoking alternative).

- Where the client is concerned about his moral values changing as a consequence of his thoughts, ask him to rate how wrong he believes stealing is, then to try and fantasize about stealing without being caught, three times a day. At a later point, have him rerate how wrong he thinks stealing is. Which theory does the evidence support—value erosion as a consequence of his thinking (a "danger problem"), or concern about this happening, without a realistic basis (a "worry problem")?

- When clients have fears about unwanted consequences of obsessions on remote events, instruct them to think about your laptop running out of battery power, breaking a kitchen appliance, or winning the lottery that weekend, and afterward process the data in terms of the relevant theories (this may uncover a malign selectivity in the power of thinking as applying to negative events only—what is the basis for that?).

OVERIMPORTANCE OF THOUGHT CONTROL

This category of appraisal tends to follow from attaching inflated importance to thoughts and tends to be featured particularly in the subtypes of aggressive, sexual, or religious obsessions (also called *repugnant-thought obsessions*). Within a schema of viewing the occurrence of a thought as signifying an alarming, credible possibility, the OCD solution is to use strategies such as thought suppression or neutralization. Paradoxically, because in OCD the aims of these thought-control strategies are not predicated on a scientific understanding of cognitive functioning, they tend to backfire spectacularly when they fail, potentially providing further discomfiting material for clients to worry about as being supportive of their fears and indicative of their personal failings.

It may seem to present an elegant solution where an overestimation of the importance of thoughts is successfully addressed using the strategies outlined in the previous section; then, the rationale for thought control may be invalidated. However, a more comprehensive solution is to address both sets of appraisals, including the erroneous significance attached to failures in mental control and thought recurrence.

Cognitive strategies. One of the simplest strategies is for the client to consider the pros and cons of holding the belief that certain thoughts or images need to be controlled. The client can list these under two columns on a page—the point is to bring home the drawbacks of the Sisyphean task of fighting against the experience of unwanted thoughts: it is time-consuming and frustrating and produces a corrupted outcome.

It may be helpful to examine whether the purported values expressed by attempts at thought control are in fact being served (this point overlaps with the previously considered category of appraisal). For example, if the value is not to blaspheme, does the experience of *unwanted, distressing automatic* thoughts designate blasphemy, or psychological disorder? If the latter, which response, other than thought-control attempts, would be commensurate with a value of rational self-care?

In light of the client's knowledge about God, how would God like the client to respond: (1) by introducing an unhelpful, ineffective, OCD-entrenching thought-control strategy, or (2) by not trying to control the thoughts and attaching little significance to them, instead acknowledging the background reality of a preference not to have the thoughts and a value of living a virtuous life?

A further point is that the research findings suggest that the client's God-created mind has a basic property of automatically exploring a wide spectrum of possible action (including undesirable action) and prioritizing the attentional focus on sources of danger, including mental experience (obsessions) that has been designated a threat (see Salkovskis & Freeston, 2001). Given these baseline facts, moral evaluation should reasonably only be introduced at the juncture where disengagement from intentional, desired "immoral thinking" or refraining from action is attainable and not counterproductive. Preferably, moral scrutiny should primarily be focused on behavior or on thoughts expressing a clear intent to act (but with action being the primary moral reference point).

Behavioral experiments. Purdon and Clark (2005) describe a catchy behavioral experiment requiring the client to construct an image of a white poodle in a bikini (?) and then visualize it for several minutes. The client is then instructed *not* to have the image for a week but record instances of its reactivation. In the next session, the discussion considers the (predictable, rather than sinister) cues that prompted the inevitable reactivation, such as seeing dogs or advertisements for holidays in which people wear bikinis, or thinking of therapy. Do the recurrences mean the client is developing a fetish for poodles, or that the images people try to suppress become more salient and harder to control?

In a similar vein, to illustrate the normal workings of the mind, instruct the client to construct and focus on an image (a green hamster?) for a minute and then not think of the hamster for a few minutes. In the first instance, she should keep track of how the mind tends to wander, and in the second instance, of having thoughts of green hamsters anyway. The experiment seeks to illustrate that much automatic, haphazard thinking goes on, with little consequence or significance, and will follow its own course, sometimes against the client's preference or intention and defeating attempts at perfect control. Ultimately, though, attentional focus on valued projects can be reestablished after temporary distraction.

A variation of this experiment, with application to OCD symptoms, is to have clients compare the interference caused by obsessions on alternate days—days when they try to control the obsession as much as possible, versus days when they completely abstain from control. Their symptoms will likely decrease on days when they abstain from control. A prerequirement for this experiment is that the client is able and willing to tolerate experiencing the obsession without ritualizing or attempts at control; to allow this, there may first have to be some progress in modification of the belief about the level of threat posed by the obsession.

PERFECTIONISM

A label of clinical perfectionism applies when the pursuit of perfection is significantly at the expense of effective functioning toward attainment of valued goals. This certainly applies to its expression in OCD. Perfectionism features across a broad range of obsessions and can manifest in a diversity of standards being pursued, such as where a to-do list has to include every single activity, the letters on the check have to be perfectly symmetrically written, or the client feels that he should always feel equally enthusiastic about his fiancée's appearance (otherwise she may not be the right one).

The overlapping aims of cognitive restructuring are

- for the client to reconsider the functionality of his perfectionistic pursuit, by reflecting on (1) the costs and goal-relevance of standards being pursued, and (2) the assessment strategy used for attainment of standards and definition of what constitutes satisfactory attainment; and
- to identify a more constructive alternative, which presents a better long-term solution in terms of his life goals.

Cognitive strategies. The functionality of standards can most simply be explored in a discussion about the pros and cons of pursuing a particular standard. As applies to most cognitive work, given the constraints of human working memory, this exercise benefits from putting pen to paper. Disadvantages of perfectionism can include, among others, that a task takes longer and becomes anxiety provoking and less enjoyable ("*have* to" subjugates "*want* to"); therefore, procrastination, overwork, and missing deadlines are frequent problems.

Exaggerated standards are defined as those of which the pursuit comes at considerable expense, thus potentially nudging the client toward achieving a significantly diminishing or even a negative return on his efforts on the inverted-U curve between productivity (vertical axis) and effort (horizontal axis)—you end up losing because of being pressed too hard. An example is when a student succeeds in writing a perfectly phrased essay with no typos but experiences high stress and short-lived satisfaction and loses marks for missing the deadline.

An issue sometimes underemphasized in work on perfectionism is the relevance of the standard being pursued to the demands of the task (i.e., is the standard relevant to the commonsense function of the behavior required to fulfil the task?). On closer reflection, absorption in the OCD can translate into *arbitrary standards* gaining preeminence over what the commonsense priority should be, such as when a client emphasizes the sound of clicking the light switch over the function of the click (switching the light on and off) or when the *exact* position of the TV remote on the table becomes more important than the function of the placement (to be easily found when needed).

The flagship perfectionism thinking error is all-or-nothing thinking, but this can combine with magnification, catastrophizing, and other thinking errors, adding insult to injury. Address all-or-nothing thinking by dividing up the client's dichotomous outcome (e.g., perfect versus unacceptable) for attainment of the standard into multiple categories, such as shocking, not great, just okay, great, and perfect (using an ordinal scale here). Or introduce a percentage scale. Using this expanded range, the client can consider whether "perfect" has to apply to *every* single situation he faces, or whether a more nuanced, flexible strategy (where in some situations "just okay" or "great" is *good enough*) would overall provide a better net return when subtracting costs from benefits. This could apply, for example, where different standards of required home neatness or cleanliness apply to different categories of visitor: the nitpicking boss you desperately want to impress versus the laid-back school friend versus the person who comes to fix the fridge, and so on.

Reflect carefully on what language the client presents for providing a rationale for her perfectionistic pursuit, so as to expose the pernicious yet seductive OCD self-cannibalizing rhetoric. For example, a client might say, "When my clothes are perfectly folded, books are organized (according to size), and nothing is lying around, we're good to go. I'm in control and ready to work." In this case, the therapist can question what *healthy control* really means: who is more in control—someone who is flexible and can respond to an urgent deadline by getting on with the job, or the person who first needs to meet a time-consuming set of standards before she can proceed with the task? In the latter instance, the reality is of the client *being controlled* (by OCD), rather than being in control.

With perfectionism, distancing strategies such as hypothetical interactions can be useful for creating cognitive dissonance with the OCD belief system. For example, you can ask the client whether he would like his children (or younger relatives, or somebody he mentors) to adopt his standards. If not, then why not? Or, ask why he may hold his friends to different standards: "So you're saying that it's okay for John occasionally to turn up a bit late—that doesn't change your view of him; but in your case, it means something worse—how does that work?" Perfectionistic standards may point to the operation of an underlying negative self-belief (see section below), which results in perfectionistic behavior being adopted as a compensatory rule-bound strategy (e.g., *I'm hopelessly ineffectual; therefore, I need to create a zero-mistake zone around me*).

It can be useful to explore the history of the perfectionistic standard ("You weren't born with this, so you must have learned it somewhere over your life—do you have any ideas about where you took this thinking habit on board?"). Sometimes the origin can be traced back to the expectations of a critical parent or other challenging circumstances. If this was the case, you can present the client with the question, "Is this really who *you* want to be now and moving into your future? Or is this an unhealthy residue of your dad's expectations that has continued into the present, is no longer helping you, and calls for urgent rethinking?"

Sometimes recurring obsessional doubting that a perfectionistic standard has in fact been attained is superimposed on the pursuit of the standard (e.g., where the client is ritualistically cleaning in response to a doubt—*Maybe it's not clean yet*—but judged by a sensory criterion it is already perfectly clean). This idea is further developed in IBT in chapter 5.

Behavioral experiments. It can be a powerful exercise for the client to experiment with pursuing a commonsensical, helpful standard in her behavior instead of complying with an OCD-maintaining, arbitrary standard or reducing the threshold she sets for defining what is acceptable performance (developing the concept of "good enough") to address OCD-exaggerated standards. For example, the client can experiment with making a spelling mistake in an email to "test-drive" the experience, test whether priority standards are compromised (will the message be misunderstood?), and test what the reaction is and whether catastrophic expectations are borne out. She can also disrupt the rigid organization of her living space every evening for a week to see if it is still possible to get work done and if feelings of discomfort diminish over time.

In a variant of an experiment described in Clark (2004), a client can contrast pursuing different 0 to 100 "perfection" ratings when performing an obsession-relevant task (e.g., 80% versus 95%—"the status quo" versus "even better"), comparing the cons (such as the time to complete the task, the effort, and level of distress) of each condition. What are the trade-offs of "boosting" the perfection rating?

INTOLERANCE OF UNCERTAINTY

Uncertainty manifests across a range of OCD clinical symptoms and is commonly also linked to other dysfunctional appraisals, such as overestimation of danger (and therefore being uncertain about safety) or perfectionism (uncertainty that a perfect standard has been attained). For example, clients describe uncertainty about whether they are gay, whether they understood something correctly, whether they can be sure that they and the world exist, and whether they really love their partner. Uncertainty appraisals are perhaps most at the fore in "need to know" obsessions, where people feel compelled to achieve certainty in respect to attaining apparently arbitrary information (e.g., "I need to know whether the neighbors' lights are still switched on before I can go to bed" or "I need to know how that building was engineered").

People with OCD may set a higher standard for concluding that safety or satisfactoriness has been achieved for uncertainty to be resolved ("I can only be certain that my hands are clean when I have seen myself scrubbing every section"), which invokes irrelevant or exaggerated criteria ("Someone else would say my hands are clean, but they don't *feel* clean to me"). In this case, the appraisals can form part of a doubting narrative that has invalidated a reasonable standard for closure, generating chronic uncertainty (further developed in IBT; see chapter 5).

Therefore, the elimination of doubt may become a prerequisite for not performing compulsions, or for terminating them, but recurrent doubting ("yes... but") makes it hard to pin down slippery certainty. Clients may introduce arbitrary ritual termination criteria (i.e., a maladaptive, noncommonsensical standard that defines the endpoint) in a desperate attempt to establish certainty ("I'm okay if I have repeated the check *five* times"). Or they may require (1) recollection of routine, high-frequency behaviors (e.g., that I switched off my computer), (2) a difficult or impossible-to-attain level of lucidity in this recollection, or (3) a clear recollection that unwanted behavior has *not* been performed (e.g., that I didn't coerce my girlfriend sexually), in order to accept that the behavior has not been performed (see MCT "Stop Signals" in chapter 6). Most frustratingly, sometimes impossible criteria are set for resolving uncertainty, such as when a person may seek the certainty of an empirically evidenced solution for resolving a metaphysical question ("Prove that the world is not a simulation").

The quest for elusive certainty is likely a driving force in most OCD ritualizing. This is motivated by poor tolerance of its dichotomously viewed antithesis, uncertainty, which is perceived as potentially leading to undesirable consequences if left unattended, such as calamity (fire, illness, being blamed) or mental turmoil (feeling chronically anxious and beleaguered by obsessions).

Cognitive strategies. Intolerance of uncertainty can be illustrated with a case example of a client employed as a researcher, who was preoccupied with needing to have a detailed understanding of every method described in a scientific paper he read as part of his literature review, in order to understand the findings. This resulted in extensive additional work devoted to this issue, culminating in missed deadlines and high stress. He sought professional help after his supervisor gave him feedback that his productivity was unacceptably low.

In this case, downward-arrow questioning uncovered layers of belief, as can be seen in figure 4.3. Pay particular attention to the intrusive thought and the appraisal.

In this case, the exaggerated (in OCD, frequently unattainable) standard underpinning the client's uncertainty that he does not know enough appears to be that he believes that he has to have an expert understanding of the statistical method used in order to be able to fully understand the conclusion of the paper to allow him to be a competent researcher and for his research to succeed. This contention as part of the OCD appraisal then becomes the focus of cognitive restructuring: first, dichotomous thinking is targeted—understanding perfectly, or not at all, versus a more flexible view that the level of knowledge required needs to be considered in light of what "job" it is required to do (e.g., Does a pilot flying an airliner need to have a detailed molecular understanding of the metals used in the wings?). Further, consider that his knowledge in this context needs to approximate that of his colleagues and supervisor, rather than that of a statistical expert: *good enough* can also be okay; he may never have certainty that his knowledge is exactly the same as theirs, but different is not necessarily better or worse.

Further questioning explored the conditions for being an incompetent scientist and doing flawed research (Therapist: "What would it take?"): being sloppy, not being methodical, not caring to correct mistakes, concealing mistakes, being generally ignorant, not seeking colleague feedback, being lazy—none of these applied to the client.

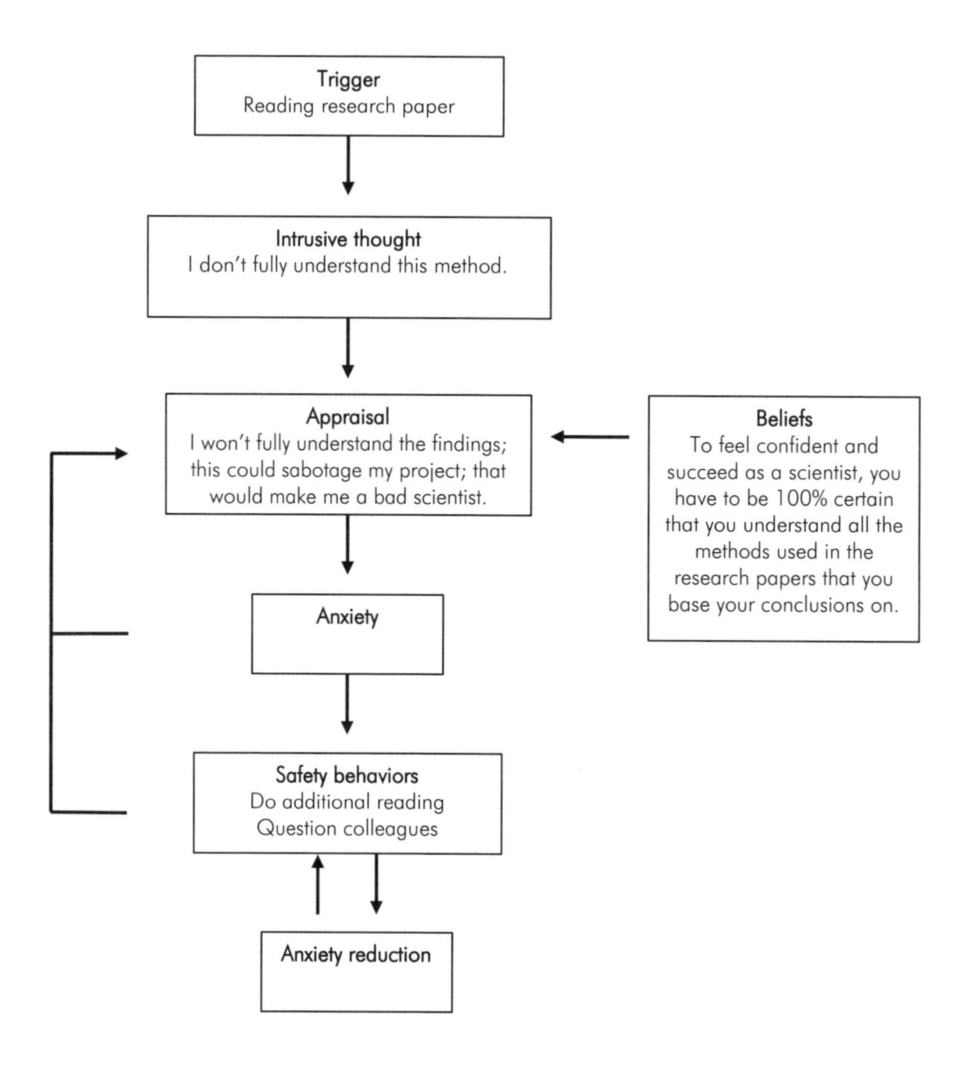

Figure 4.3. Intolerance of uncertainty/perfectionism case conceptualization

Finally, the therapist invited the client to list the pros and cons of his striving for certainty about having sufficient knowledge, and after that to question the validity of the pros (e.g., Does striving for certainty in this situation actually benefit performance? What are the trade-offs? Are the priority attainments being served or impeded?). Next, they weighed any pros (better understanding of statistics) against the cons (that striving for certainty may be a never-ending quest, sabotaging job satisfaction by causing worry and anxiety, impacting negatively on the client's work relationships and resulting in compromised productivity…so more ends up being less).

In this case, the aforementioned cognitive work led to behavioral experimentation with ritual prevention (see "Behavioral experiments" below)—the client ceased doing additional reading and repeated questioning of his colleagues. Aligning behavior with more realistic goal pursuit allowed him to realize that feared outcomes did not transpire—the intensity of thoughts of uncertainty and anxious feelings subsided over time, and in fact his project benefited from working in a more focused way. This and positive feedback from his supervisor allowed the alternative appraisal to be strengthened further, the power of the obsessions to diminish, and the positive behavioral trajectory to be sustained.

A further example of where it may be helpful for clients to reconsider the criterion applied for certainty versus uncertainty in their appraisal is obsessions about sexual orientation. For example, if the client fears being gay on the basis of a thought that she may have been sexually attracted to someone of the same sex, you can present a continuum model of sexual orientation, with homo- and heterosexuality as the ends and most people falling somewhere on the line other than being exclusively one or the other. (Reading assignments about sexuality can be useful here.) In light of this nuanced view encapsulating the complexity of sexual experience, ask the client, "What are the pros and cons of pursuing a definitive answer, rather than 'just letting the question sit there?'" Is there a danger of overgeneralizing on the basis of "outlier" experiences and ignoring the bulk of relevant lifetime experience?

Continua can also be helpful with "need to know" obsessions, presenting clients with a strategy for grading and relativizing the importance of needing to know on a scale of 0 to 100, such as the favorite band of the checkout person versus the first name of the new intern versus knowing that you have hypertension. This should prompt consideration of which are reasonable and helpful criteria that should determine the importance of knowing information.

In my experience of clients ruminating about metaphysical questions (e.g., *Does the world really exist?*), it can be useful to point out that they are not pursuing the question in the usual meaningful way (for those with a philosophical interest), as life-enriching philosophical pondering. Rather, their ritualizing is driven by their fears (e.g., *If nothing existed, I'd lose everything; I'd be all alone*), and the criterion for certainty is unattainable (such as resolving the question empirically, formulating a definitive philosophical "proof," or getting an enduring *feeling* of closure). Which would be a simpler account—that the world exists, or that it is a simulation? How can this metaphysical question appropriately be "resolved" within the parameters of its category? Can it be proved right or wrong? What are the pros and cons of striving for certainty?

It may be helpful to reflect on the advantages of parsimonious explanations (i.e., why simpler is better). An analogy can be useful: if I am walking in the woods by a pond and see a puddle of green-brown goo on the ground, it could be explained as being pond sludge, or the poisonous dropping of a frenzied extraterrestrial who earlier passed by. Which is the preferred explanation and why? (Be careful, though, to present this topic in a nonpatronizing way.)

Also, contrast the ruminative existential questioning with the client's acceptance of things as they are in other domains of life—why does she torment herself about *one* existential question while facing so many others with equanimity? What useful information can be gleaned from the client's composure about similar questions?

Behavioral experiments. As usual, the most useful experiments allow testing of clients' predictions about what might go wrong, or how bad the experience might be, if they terminated their ritualizing. For example, clients may predict that the intensity of needing to know will continue undiminished if they do not ritualize, and that this will severely impair their functioning. They can experiment with not acting on their obsession and then tracking over time their anxiety and relevant markers of the level of interference in their functioning (e.g., being able to attend to non-obsession issues); the likely outcome is that their obsessions and associated unease and any impediment fade and dwindle over time.

CONSEQUENCES OF ANXIETY

Typical beliefs in this category concern the prediction that anxiety is dangerous and harmful and that the client may be overwhelmed by it or lose control.

Cognitive strategies. When considering the evidence relevant to fearful predictions, the following may be useful:

- Provide psychoeducation: reflect on the function of anxiety and its physiological manifestation—preparation for fight or flight, aimed at *increasing the chance of survival* rather than presenting a source of danger.
- Reflect on the overlap in the physical experience of anxiety and doing exercise: How well did the client manage to tolerate the sensations associated with exercise? What does that say about the dangerousness of anxiety-associated physiological activation?
- "Unpack" vague labels used to describe catastrophic outcomes in terms of their behavioral markers: What does "losing control" really mean? Has the client experienced this before? If not, what does that mean?

Behavioral experiments. Exposure and response prevention experiments are the most pertinent strategy here. The client tests *specific predictions* about the feared consequences associated with being anxious following exposure to obsession triggers, such as her ability to function behaviorally and mentally ("I won't be able to concentrate on anything") and duration of peak anxiety ("I'll be 100% anxious for hours on end"). Operationalize, or make measurable, the client's predictions (e.g., "not being able to concentrate" means that she won't be able to read and comprehend a short newspaper report). Next, carry out rigorous testing and comparison of the predictions with obtained results, and reflect on what any discrepancy means.

Core Beliefs

Client core beliefs may become evident as more general themes in the client's situation-specific automatic thoughts emerge and as deeper layers of thinking are excavated when doing downward-arrow questioning (e.g., *I am bad/dangerous/abnormal/unlovable/ weak/vulnerable/out of control/incompetent/worthless*). Working on emotionally resonant negative core beliefs requires good rapport and a trusting therapy relationship and tends to be appropriate mainly later in therapy. Standard CBT strategies apply, also in the context of OCD treatment:

- Psychoeducation: introduce the client to an information-processing model, and explain the core belief as a funneling structure that "streamlines" and prioritizes the processing of information aligned with the belief, while inhibiting and distorting the processing of information that contradicts the belief. (Therapist: "You end up seeing only what you're already believing, instead of believing what you're seeing.")
- Standard CBT weighing up of the evidence for and against the core belief is a legitimate first line strategy, followed by formulating an alternative belief to give a fairer account of the full breadth of data.
- A proactive strategy for "constructing" a new adaptive belief is to have the client schedule behaviors consistent with the new (desired) belief on a daily basis (e.g., *Because I am basically competent and doing an okay job, I can afford to take a proper lunch break*) and practice being on the lookout for and acknowledging evidence over the course of the day that supports the new belief. This has the advantage of embodying the new belief in recurrent behavior and reinforcing it through ongoing processing of consistent data.
- If childhood events were powerfully instrumental in the formation of the core belief, the client can consider each in turn and introduce a remedial, "healthy adult" perspective to update the distorted conclusions reached during the turmoil of the traumatic past (e.g., the client can

imagine being in the situation and thinking of his own adult self or some other trusted figure entering the situation and offering a corrective viewpoint).

The reader is referred to the excellent texts by Beck (2011), Wilhelm and Steketee (2006), and Young, Klosko, and Weishaar (2003) for more detailed discussion of core belief work.

Relapse Prevention

In the later stages of treatment, the aim is to encourage clients to assume greater autonomy in therapy, acknowledge the contribution of their own efforts to progress, and hone their skills in order to be able to apply self-therapy skills independently.

This is an appropriate time for detailed reviewing of which CT techniques and insights were most valuable in therapy and how these could be applied in the context of early warning signs of relapse or in the face of novel stresses the client may encounter in future. Other topics worth considering are support-group involvement, use of medication, and how clients can fill the time no longer occupied by their OCD symptoms with meaningful, resilience-enhancing activities (e.g., hobbies, socializing, structured exercise).

To allow a comparison with the baseline assessment, administer questionnaire measures mid-therapy or at least by the final session to identify any final issues that need to be addressed. Hereafter I prefer to schedule at least one or two follow-up sessions, separated by intervals of a few months, to consolidate therapy gains over a longer time frame.

CHAPTER 5

Inference-Based Therapy (IBT)

Inference-based therapy (IBT; O'Connor, Aardema, & Pélissier, 2005; O'Connor, Koszegi, Aardema, van Niekerk, & Taillon, 2009) was developed by Montreal-based clinical psychologists Kieron O'Connor, Fred Aardema, and Marie-Claude Pélissier. You'll also see reference in the literature to the inference-based approach, or IBA, which refers to the theoretical model on which the therapy, IBT, is based; for simplicity sake, I will refer only to IBT here. If the "C" in CBT and the study of rhetoric and reasoning float your boat, this imaginative approach will likely be of particular interest to you.

IBT was originally developed for the difficult-to-treat category of OCD with overvalued ideation. There is preliminary evidence of a possible advantage of IBT over standard CBT in this group (see Julien et al., 2016); however, IBT has been found to have application across most types of obsessions.

Below, I will consider key concepts in this approach before introducing the clinical model. Finally, I will describe an abbreviated IBT protocol (see van Niekerk, 2009; van Niekerk, Brown, Aardema, & O'Connor, 2014) with optional components for integration with the strategies of ERP and appraisal-based CBT, covered in chapters 3 and 4. However, IBT is the underpinning framework and constitutes the bulk of this protocol. For a description of the full IBT protocol, I can recommend the excellent clinician's handbook by O'Connor and Aardema (2012). This and the other references provided serve as the source material for this chapter.

Key Concepts

IBT focuses on the role of reasoning in OCD. It differs from the cognitive appraisal model (chapter 4) in that it does not consider obsessions to have their origin in the appraisals of normal intrusive thoughts. Instead, no matter what type of OCD, the clinical kicking-off point in the obsessive-compulsive chain is a conclusion of doubt arrived at through a dysfunctional reasoning process characterized by what IBT calls *inferential confusion*. The backdrop of people's narratives (or stories) about themselves and their lives ferments the triggering of obsessional doubts in different situations as a consequence of the style of reasoning they adopt. These concepts will be further developed below.

Doubt

The concept of doubt is central in IBT, but what does it mean? Doubting always involves questioning the information the person already possesses. You doubt what you *know*, which is how doubting differs from genuine uncertainty, which entails a lack of information. To illustrate the difference, consider these examples: (a) it's sunny and I walk inside; I hear raindrops, leading me to doubt that the sky is still clear (doubt), versus (b) I am asked what the weather will be like tomorrow, and I haven't checked the weather prediction (uncertainty)—I simply lack the data.

How does doubting manifest in OCD? Obsessional thinking always entails a reevaluation of information the person already has, suggesting that an object, situation, experience, or the self has been compromised and is no longer satisfactory or okay. As a result, the person reaches an unsettling conclusion of doubt, such as that the door handle could be dirty (*even though it looks clean*), that the tap could still be on (*even though I saw myself turning it off*), or that I could lash out impulsively (*even though I have never done this before*).

The Feared Consequences

Once the client is absorbed in the doubting narrative and considers it credible, it is logical to think about what bad events would follow if the person did not try to prevent harm or set things right by performing safety behaviors (e.g., ritualizing, thought suppression). Following from the above examples, the responses might be "I could infect my child" (*if* there were germs and I didn't wash), "The house could get flooded and it will cost a lot of money" (*if* the tap was left on and I don't go and check again), and "I'll get thrown into prison and I'll be consumed by guilt for the rest of my life" (*if* I lashed out, which could have been stopped if I blocked the thought and removed myself from the situation).

Once the person starts contemplating the feared consequences and imagining being faced with such a situation, negative emotion enters the fray, strengthening the obsessional narrative—the sense that something is wrong and remedial action is urgently called for.

Primary vs. Secondary Inferences

In IBT, the doubt is sometimes referred to as the *primary inference* and the feared consequences as the *secondary inferences.* An inference is simply an opinion that you form on the basis of the information you have (i.e., which you reach through a process of inferential reasoning). Primary versus secondary here signifies that the doubt is the premise that applies *before* what is implied—the feared consequences—can be considered.

Inferential Confusion

Doubting is a common experience in life: when sitting in a restaurant, the "juicy" T-bone steak on the menu sounds wonderful—until you see someone else's scrawny excuse for a T-bone being served at the table next to you. Your initial premise (based on the knowledge you had) was that the steak was a good choice. You then acquire new information (seeing the steak served), on the basis of which you reason to invalidate your initial premise—maybe the steak isn't such a good idea after all. Notice here and in the example above, *new information about reality was acquired from the senses* (hearing rain and seeing the steak), and this led to doubting. Therefore, the cognitive ability of *perception* (the process of recognizing and interpreting sensory stimuli) led the way in establishing the basis for the doubt.

However, in OCD, obsessional doubts are inferred on the basis of an invalid and irrelevant reasoning process, which reveals evidence of *inferential confusion.* Inferential confusion has two parts: (1) an investment in remote possibilities in preference to reality, and (2) a distrust of the senses, or the self, and common sense ("common sense" can broadly be defined as that approach that promotes constructive behavior, helps people attain their life goals, and is justified by the requirements of the task at hand; see O'Connor et al., 2005; O'Connor et al., 2009).

These two components result in the person with OCD—against the reality of the situation—being confused into believing there is a concern that needs to be addressed, when in reality nothing is awry. For example, consider this dialogue:

Client: (*when seeing a clean door handle*) There could be germs on it.

Therapist: Why do you believe that?

Client: People sneeze and cough into their hands. They could have touched the handle and transferred germs; I heard there's a cold going around.

Therapist: But it looks clean?

Client: Sure, but you can't see the germs.

Note that in contrast to the previous examples of "normal" doubting, here the doubting was *not prompted by new, relevant perceptual information in the situation*—the handle looked clean, and the client did not see someone sneeze into his hand and touch the handle. The doubt came from within, rather than following from direct observation of something outside (the sensory reality).

The Doubting Narrative (or "OCD Story")

The doubting narrative constitutes all the interwoven reasons why the person considers the obsessional doubt to be credible when it is activated. The hallmark is the influence given to "maybes" or hypothetical possibilities. The narrative replaces confidence in the senses and the self with a doubting inference based on remote possibilities, and the person is led astray from a commonsensical viewpoint in favor of prioritizing irrelevant and unrealistic criteria.

In the example above, the possibility that someone with a cold could have sneezed or coughed into his hand, touched the door handle, and transferred germs trumps the reality of simple perception. The narrative bridges the gap between sense information (a clean handle) and the obsessional inference of doubt: that the handle could be contaminated (see figure 5.1).

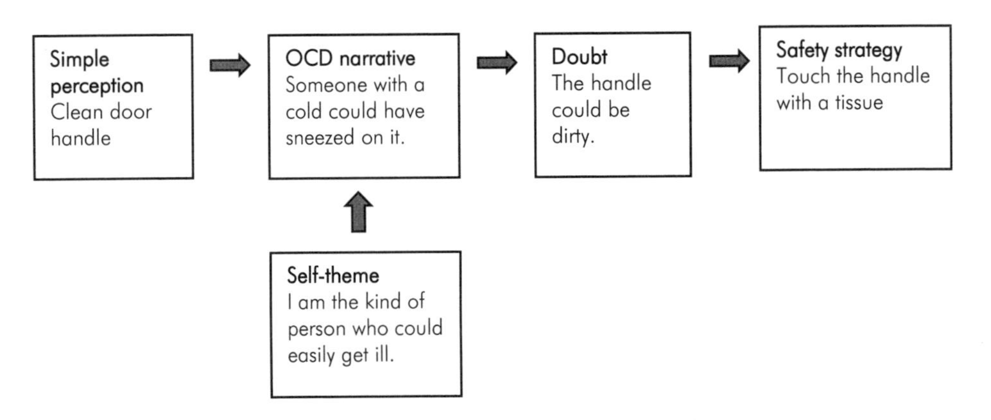

Figure 5.1. Narrative bridging between perception and doubt

The emotional aversion associated with the envisaged consequence of touching the handle—being infected—and the consequent behavioral response (touching the handle with a tissue) further validate and strengthen the doubt. This happens because there is intimate cross talk between emotion, physical sensation, behavior, and thinking (see the "somatic marker hypothesis" discussed in Damasio, 2005). Further, acting as if the doubt is valid constitutes behavioral rehearsal. "Testing behavior" tends to backfire, as when a client fears unwanted sexual attraction and tests this by closely focusing on his genitals while he engages with material that he usually avoids; the perceived sensation caused by the shift in attentional focus is then misleadingly interpreted as validating the doubt, which requires further anxious verification, and so on.

The OCD story derives its power to kindle doubt from the characteristics of its narrative architecture. Humans love listening to (and telling) good stories! For example, in a classic study by Heider and Simmel (1944) in which students were shown a film clip depicting movements of simple geometric shapes, when asked what they saw, all but one of the thirty-four students described animate beings, such as a tale about "two innocent young things" represented by a triangle and a circle! A more convincing argument or captivating story tends to be one that resonates with us emotionally and is rich in detail (think of a TV series such as *The Wire* or good films or literature). When reading a good book, you're transported into a different world (I still recall with trepidation entering the world of Stephen King's *Pet Sematary* as a teenager…and yes, it is spelled this way!).

The OCD story is a deeply personal story that contains a number of rhetorical devices that amplify its believability (rhetoric is the art of persuasive speaking or writing). IBT calls them devices rather than errors because in some contexts they are appropriate and helpful (e.g., arguably, used by Stephen King when writing *Pet Sematary*). However, in the OCD situation, they transport the person *away* from reality, gliding along to a dark and disconcerting land on the powerful magical carpet of the human imagination, the cognitive faculty that deals with possibility. (The imagination deals with what *could be there*, whereas perception deals with what *is there*.) A detailed description of individual thinking devices will follow below.

The Vulnerable-Self-Theme

In most OCD cases, situational obsessions are propped up by an underlying, all-encompassing vulnerable-self-theme (the feared self), with roots in the person's history, about who the person may be or may become (see Aardema & O'Connor, 2007). These fears apply to selective life domains, as apparent from the selective focus of the concerns represented in the obsessions. However, this view of the feared self runs directly contrary to the reality of who the person is (the authentic self), or any realistic basis for who the person may become.

The broader self-theme can usually easily be inferred backward from individual obsessions, such as being *the kind of person* who could be especially at risk of...getting contaminated/leaving the door unlocked/swearing impulsively in church/making preventable mistakes, and so on. The IBT model holds that the hypothetical feared self underpins doubting across diverse obsession-triggering situations, such as when believing oneself to be "someone who is careless/a scatterbrain/an accident waiting to happen" manifests in obsessional fears about making disastrous mistakes in diverse situations.

Similar to situation-specific obsessional doubts, the self-theme is approached as a conclusion that derives from a self-doubting story. The narrative can be uncovered by carefully questioning the client about why she considers the view of the feared self to be credible. This reasoning will reveal elements of inferential confusion, similar to individual obsessions, which treatment will seek to address.

The aims of IBT are to invalidate the reasoning producing obsessional doubts and return clients to the world of the senses and common sense, which they have been led to disbelieve and mistrust. Clients are helped to gain metacognitive insight into the nature and origins of the obsessional stories—they learn to step back and realize they have been overlooking reality, instead of going deeper into reality. This helps them realign their thinking and action in the obsessional situation with their behavior in the situation before OCD gained a foothold—or a stranglehold—and with how they interact with non-OCD areas of their lives.

Clinical Model

IBT concepts (see figure 5.2, adapted from O'Connor et al., 2005) come together in a symptomatic situation in the following way: First, the person experiences an internal (e.g., a thought or a memory) or external (e.g., an object) *trigger*, which activates the *obsessional doubt*, the product of an explicit at-the-fore, or implicit in-the-background, doubting narrative (the OCD story). The doubt is represented by the first point where the person goes beyond the immediate reality data and strays into the imagination.

The person envisages negative consequences (the *feared consequences*) if the doubt were true and he did not act. This results in the activation of *negative emotions* (e.g., anxiety, guilt, discomfort). The person is motivated to perform a *safety behavior* (e.g., ritualizing, reassurance-seeking), which may have to be performed repeatedly as part of a ping-pong match with rebound doubt seeking to undermine the attainment of closure. Given that the person has invalidated direct feedback from reality, it is easy for the imagination to sidestep reality, manifesting as "yes...buts" when reflecting on a reassuring viewpoint; frequently an arbitrary ritual end point is selected (e.g., "I have done it five times" or "It feels right"). Of course, the person may decide to avoid the situation altogether, in anticipation of doubting-related distress, hassle, and impediment.

As described before, IBT primarily seeks to intervene in the doubt and the OCD story, seeking cognitive change that will make the use of safety strategies pointless. However, these goals are not incompatible with the use of methods as part of ERP or cognitive therapy. An integrated approach using standard cognitive therapy techniques can also target distorted or catastrophic thinking about the envisaged feared consequences, for weakening the emotional momentum toward ritualizing. At the final base, structured response prevention work can contribute to eliminating safety behaviors. Below, I will describe an IBT treatment protocol with supplementary standard cognitive therapy and ERP components.

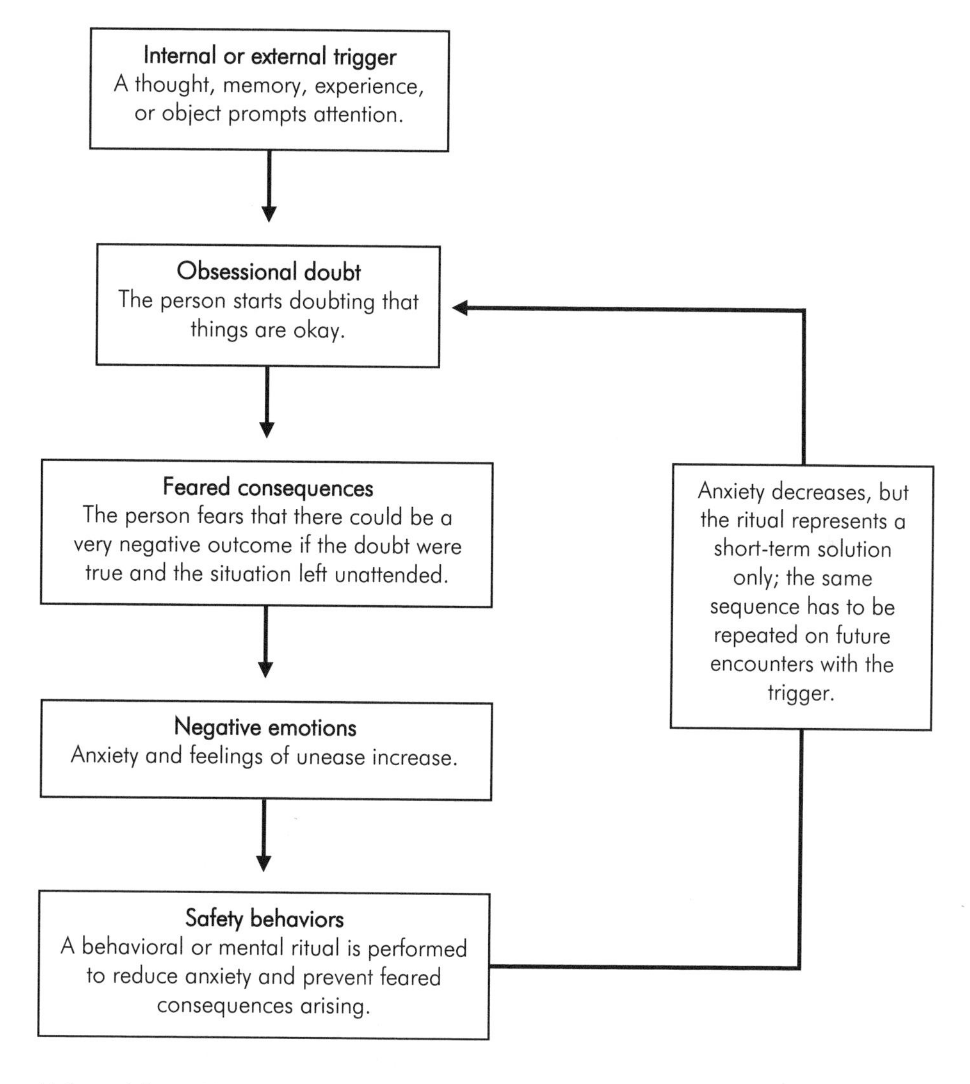

(Adapted from O'Connor, Aardema, and Pélissier, 2005)

Figure 5.2. The inference-based therapy model of obsessive-compulsive disorder

IBT for OCD

The treatment protocol typically requires fifteen to twenty weekly, one-hour sessions (although there is some flexibility with spacing of sessions), excluding follow-up. Building on the initial assessment described in chapter 2, more focused IBT assessment follows, which informs the case conceptualization and intervention. These are considered below. I routinely give clients homework assignments to read chapters of my self-help book (van Niekerk, 2009), not as preparation, but to consolidate therapy insights, therefore preserving a format of *guided discovery*; a wealth of IBT client materials and other information is also available at http://www.ibaocs.com.

Assessment and Case Conceptualization

First, list the main obsession-eliciting situations and conceptualize them according to the IBT model (figure 5.2), allowing identification of the trigger, the doubt, the feared consequences, and the safety strategies for each situation. Not every obsession has to be listed, but the main obsessions, in terms of prominence and severity, should be represented. It is useful for the client to keep a basic symptom diary (appendix 1; also available at http://www.newharbinger.com/38952) for a week, to inform this part of the assessment.

The following questions can assist in identifying the obsessional doubt (and any linked negative meanings) in the main trigger situations:

- What do you think is not okay in the situation?
- What is the ritual intended to change?
- What do you have to be 100% sure of not to perform the ritual?
- Why perform this ritual and not another one?

If, in response to the above questions, the person describes the feared consequences, ask her what would have to apply for the feared consequences to occur. Remember, the doubt is the first point where the person leaves reality.

Therapist: What troubled you in this situation?

Client: The house could be burgled.

Therapist: And what would make that possible?

Client: If I left the window open.

In this example, the doubt is *I could have left the window open.*

To identify the feared consequences, ask:

- What do you worry may happen if you didn't do the ritual and the doubt were true?
- If you did nothing, what could go wrong here?

To identify the emotions, ask:

- Which feelings did you experience when you thought about what could go wrong here?
- Which feelings was [describe the ritual] meant to change?

Finally, to identify the rituals or safety strategies, ask:

- What did you do to reassure yourself, set things right, or reduce your anxiety?

If the client is coping by avoiding the trigger situation altogether, simply ask the questions above as if the client had actually encountered the situation.

These are abbreviated examples of IBT case conceptualization:

Doubt(s)	Feared consequences	Safety strategies
There could be herpes virus on the towel rail.	I could be infected.	Don't touch it.
The iron is still on.	The house could catch fire.	Check the iron.
I could have been sexually attracted to the child—I could be a pedophile.	I couldn't live with myself.	Seek reassurance from OCD online forum.
The light switch didn't click in the right way.	I'll feel on edge all day long.	Keep clicking.
It's my moral duty to pick up the plastic bag.	I'd keep on worrying about being to blame if a child was smothered by it.	Pick up the bag.
Maybe it will bug me forever if I don't know what they said (others talking nearby).	My holiday will be spoiled.	Try to overhear the conversation.

Simple ratings can be taken at baseline and after introduction of the main therapy components to allow tracking of treatment progress. Ask clients to provide 0 to 100 ratings for how much they believe the doubt, how realistic the feared consequences are, and how strongly they feel the need to perform the ritual (ratings can be taken for how the client expects to feel in the trigger situation). The thirty-item Inferential Confusion Questionnaire (ICQ-EV; Aardema et al., 2010) measures inferential confusion in a simple way and can be helpful in guiding the focus of assessment and providing a measure of treatment mechanism (it is an appendix to the cited article and can be downloaded from the Internet).

Invalidating the Doubting Narrative

The first step in invalidating the doubting narrative is to select a doubt experienced in one trigger situation, central to the client's concerns, for intensive IBT analysis and intervention. A good outcome here would be defined by a generalization of insights to obsessions experienced in other situations or for further therapy work in these areas to be expedited. Follow the steps below:

IDENTIFY THE OCD STORY

Ascertain the reasoning that supports the conclusion of doubt by asking the person to identify all the reasons why there was a real chance the doubt was true or why the doubt was credible or convincing (this is commonly also a written homework assignment, encouraging the client to reflect patiently). Encourage the person to "let the OCD speak" and not self-censor, therefore reducing the chance of obstruction by social desirability bias (the client saying what will be viewed favorably by others) or self-criticism. Prompt the client to consider a role of self-doubting (see "The Vulnerable-Self-Theme" above): "Is there any contribution here of thinking about *the kind of person* you may be, such as someone who is especially at risk of…?"

When relating the OCD reasoning, clients typically interject something like "I know it's complete rubbish." To deal with this, you can introduce a courtroom analogy: "That sounds like the commonsense lawyer, but now is the OCD lawyer's turn—let's hear what he has to say." It is important to capture the "OCD rhetoric," so give the client ample opportunity to describe the concerns *in his own words* and avoid putting words in his mouth ("OCD is a conman, so we need to understand the tricks of his game—give him a chance to speak"). If you volunteer a certain phrasing, be careful to check with the client that this accurately encompasses his view; also ask how he would phrase it in his own words.

Occasionally clients say that there is no thinking behind the doubt. This may reflect that the ritual has become highly automatized (habitual) over time (some of the initial reasoning may have receded), or reflect a contribution of past socialization that obsessions are senseless thoughts that are the product of an abnormal brain. Or, it may be due to obstructions such as social desirability bias or self-criticism, described above. In this case, urge the client to "take [her] time and give the OCD a chance to talk" (it is very helpful to *slow down* the reasoning and record it in writing, although clients may find this uncomfortable), or prompt the client to reflect on the substance of the case for performing the ritual.

It may be helpful to consider whether there was a narrative at a previous point in the OCD: "How did the OCD talk to you about this at the time when the obsession started?" You might also encourage the client to reflect on implicit assumptions: "Sounds like the conman is saying *this* (e.g., losing control over behavior) is possible—is he also saying how that would work?" I can recall a client insisting that he did not believe the doubt at all (that actions he repeated a lucky number of times protected loved ones), but at a later point he qualified this by maintaining that he still did not believe it but "could not take the chance that it wasn't true." The OCD story may also reveal itself as "yes…but" as part of an internal debate; these need to be recorded as part of the narrative.

REFLECT ON THE ORIGIN AND PITFALLS OF THE OCD STORY

At this point, engage the client in reflecting on what is distinctive about the OCD reasoning—how it deviates from his approach before OCD started and how it is different from his thinking in other life areas. Aim to uncover the inception of the OCD story in the imagination (how what "*could be* there" is made to be more important and relevant than "what *is* there") or how it subverts common sense (e.g., the commonsense purpose of an activity is lost in favor of pursuing arbitrary goals).

Therapeutic leverage is offered by exploring the selectivity of the OCD reasoning. Despite some clients' insisting that they obsess in all situations, this is never the case. In the majority of life situations, they consolidate reality with common sense rather than embroidering on discomfiting, hypothetical, remote possibilities (of which an infinite number is possible in *any situation*—yes, the therapist could be an extraterrestrial?) that are being allowed to control their actions. A simple illustrative example may be when the client crosses the road: "You just accept what you see and hear (reality sensing), in a flash, and then cross." In contrast, in the OCD situation, the client can reflect on how when she gets "the OCD feeling," she ventures into the imagination ("goes into the bubble") where nothing is clear any more ("a shadowland"). You can also use the analogy of a bridge: turning away from reality on one side and crossing the bridge into the imagination ("OCD land").

This is different from the client's approach in other situations (including when the obsession is weaker) where she chooses to stay in reality and not doubt. Prompt the client to reflect on this selectivity by saying, for example, "You seem to tell stories about germs on objects, but not about germs in the air" or "You worry that you acted offensively to people in your building, but not to my receptionist—how does your thinking differ in these two situations?" If the client answered that the OCD situation is different because she had a spontaneous thought about the unwanted action, this becomes the first point to record as part of the OCD story for that obsession.

You can reflect on the costly consequence of immersion in the narrative—having OCD. However, in this and other lines of questioning, the Socratic method is recommended, respectfully normalizing the client's experience where appropriate ("We all do this sometimes"), rather than vigorously pursuing a point or trying to persuade.

CONSIDER OCD THINKING DEVICES

Next, explore the role of the OCD thinking devices (or "con tricks"), including the following (not mutually exclusive) key thinking devices (O'Connor et al., 2005):

- Making unsupported links between different categories of information, where X transmogrifies into Y, or making a conclusion of bogus equivalence or spurious causal connection ("A year ago I read in the paper that a woman swallowed glass in her cereal, so there could be glass in my cereal," or "I thought about the plane having engine failure, which could cause it to happen"), or conflating an imagined experience with a real memory, without a basis in immediate reality for such a link.
- Introducing facts inappropriately, without evidence or a plausible precipitant in reality ("Hackers steal data, so my data could have been stolen.")
- Creating imaginary stories ("Three is my lucky number because I was born on the third of March, so if I do everything in threes, bad events will be avoided.")
- Reaching upside-down conclusions (sometimes referred to as *inverse inference*), where the conclusion precedes and trumps observation ("Toilets are dirty because many people are unhygienic, so this toilet could be dirty even though it looks clean.")
- Distrusting the senses ("Even though I saw myself turning off the switch, maybe I didn't look carefully enough.")

In addition to the thinking devices proposed by O'Connor et al. (2005), there is also scope for considering the following:

- Unrealistic demands about the ability to control thoughts or about the kinds of thoughts one has ("If I can't stop an unwanted sexual thought, it means I must want to act on it.") (Frost & Steketee, 2002)
- Unrealistic expectations about memory confidence, vividness and detail, which *decrease* with repeated checking even though memory accuracy is unaffected ("I've gone over the thought of locking the door many times, but I don't feel confident that it is locked.") (Dek, van den Hout, Engelhard, Giele, & Cath, 2015; Radomsky, Gilchrist, & Dussault, 2006; van den Hout & Kindt, 2003)
- Unrealistic expectations about perceptual certainty, which *decreases* after prolonged visual fixation on an object that results in dissociative feelings ("I keep on staring at the light switch to make sure it's switched off, but it starts looking fuzzy." You can test this by staring for a minute at the word "fuzzy" in the text.) (van den Hout et al., 2009)
- Creating arbitrary or exaggerated standards, frequently at the expense of the commonsensical *function* of the object or action ("When I click the switch, it has to make the right sound.") (van Niekerk, 2009)
- Emotional reasoning ("I've got this creepy feeling, which means a window could have been left open even though I know I checked.") (see Arntz, Rauner, & van den Hout, 1995)

Identification of specific thinking devices is less important than understanding the broader principles, such as that specific remote possibilities are selectively made to seem imminent and real, that they are unsupported by sensory reality, or that arbitrary criteria are prioritized at the expense of longer-term goals. The point is that the OCD reasoning in the situation *is not necessarily wrong, but irrelevant and counterproductive.*

DEVELOP AN ALTERNATIVE NARRATIVE

The client can now progress to developing an alternative narrative (which I call the "commonsense view") to the OCD story, grounded in reality and common sense, supporting the conclusion that the OCD reasoning is irrelevant, the obsession therefore unfounded, and safety behaviors not required. This gives an opportunity to develop counterpoints to the OCD "con tricks" and "yes… buts," which can be assigned as a homework assignment.

Caveats need to be heeded about pitfalls, such as cognitive work blending into mental ritualizing (which is a danger in any OCD cognitive intervention) or trying to "prove" the (difficult or impossible to falsify) OCD story wrong. Developing the alternative story needs to be time-limited, as further doubts are likely to be activated (e.g., "Maybe I haven't done enough yet"); these are dealt with by referring back to the model (an arbitrary standard of "not feeling right" might have been invoked in support of this inference). The aim ultimately is for clients to stay on the "reality side" of the bridge when they get the "OCD feeling" and form (and act on—see below) a simple, reality-driven conclusion, *quickly*, without unwarranted pondering or debate.

ACT ON REALITY

If the cognitive work described above were effective, the client's conviction in the doubt would have diminished (as reflected in a reduction in his 0–100 ratings), as he can now access a reality-driven perspective and have insight into the pitfalls of the OCD reasoning. He can now be encouraged to "act on reality" in the situation (i.e., substitute safety behaviors with normal, adaptive behavior), using either a *one-step approach* (preferable) or a *graded approach*. (If the client is unwilling to proceed with response prevention, additional cognitive work may be required; however, beware that cognitive work evolves into ritualizing; see above.) It may be useful as a preliminary step for the client either in session or as a homework assignment to carefully read the alternative perspective and visualize responding to the trigger situation from that viewpoint (however, eventually this step has to be eliminated, so as not to transition into a safety behavior motivated by secondary doubting—*Maybe I need to do this every time*).

The one-step approach involves eliminating all safety behaviors in a single step when engaging with the situation, while an optional graded approach entails gradual elimination of ritualizing, similar to response prevention work in ERP (chapter 3). With either approach you can acknowledge a role of fear conditioning but emphasize that behavior change is "for practicing acting on reality and common sense in the normal and natural way and getting back into the habit of doing this, similar to how you think and act in non-OCD areas of your life."

Optional Module: Cognitive Interventions for Thinking About the Feared Consequences

Here, the focus shifts to thinking about the feared consequences (what might happen) if the doubt were true and the ritual not performed. *This is an optional step and not part of IBT.* It may offer helpful additional leverage if the initial cognitive components were not sufficient to allow the client to progress to ritual elimination, and if there is significant inaccuracy or dysfunction evident in the thinking about the feared consequences, which amplifies overall distress.

In this case, help the client reflect on and then answer, in a realistic and helpful way, any inaccurate thinking about what might go wrong by considering a series of questions:

1. *What do reality and common sense tell me about the likelihood that these consequences will occur?* Among other strategies as part of cognitive restructuring, you might offer an accurate method for calculating probabilities (see van Oppen & Arntz, 1994), described in chapter 4.

2. *Am I underestimating my or others' ability to cope with the consequences?* Guide the person to consider the relevant evidence and to develop an accurate appraisal of her own and others' competencies and resources.

3. *Am I taking more responsibility for the bad consequences than I need to?* The role of inflated responsibility in OCD was highlighted in chapter 4. In the current protocol, responsibility is conceptualized as relating to "your sense of which actions are reasonably required of you to prevent something bad from happening to yourself or others." Encourage the person to reflect on how much she would be to blame if the feared consequences occurred, by examining various realistic criteria for assigning a higher level of blame:

    - You intended to cause harm.
    - You were specifically assigned the responsibility of performing certain actions to prevent harm (such as being the caretaker of a children's playground) and neglected to do your duty (you didn't fill up a large, concealed hole).
    - There was a realistic chance of the feared consequences arising if the situation were left unattended (there was a clear risk that a child could break a leg, or worse).
    - There were clear and immediate signs of risk or danger (there were children getting ready to play).
    - You were the sole person whose actions contributed to the consequences (you were the only supervising adult).

    Clients may see themselves as accountable for *any* negative consequence that results from their actions (e.g., "It would be my fault if I walked past a rusty piece of wire in the road and a child stepped onto it and developed blood poisoning"). In this case, it may be helpful for the client to reflect on the difference between the responsibility for an *action* (or for failing to perform it) and responsibility for the *consequences* of an action (or of failing to perform it). The former can be considered as dichotomous, and the latter as being on a continuum judged by the factors listed above. For example, you might be responsible for dropping a sheet of paper on the pavement, but not reasonably be blamed for it blowing onto a highway and causing an accident, which would be considered a freak event! The pie chart technique (see chapter 4) can be helpful for learning to reasonably allocate responsibility. Prompt the client to reflect on society's encouragement, and the benefits to all, of living vigorously, acting when there is *real* risk or danger rather than trying to safeguard selective, hypothetical harm.

4. *Am I being unfairly hard on myself in terms of what it would mean about me as a person if the worst happened?* Encourage the person not to self-recriminate unfairly and harshly if feared consequences happened and she was in fact implicated.

Now repeat the steps detailed in the IBT modules above for other obsession-triggering situations. The usual pattern is for detailed and elaborate cognitive work on initial obsessions, after which this becomes expedited as the client becomes better acquainted with IBT principles, with a generalizing effect of earlier progress.

Self-Doubting

Elements of self-doubting (revealing the "feared self") may manifest in the OCD stories specific to individual situations, allowing you to engage with these using the strategies above to access a view of the "authentic self." However, at a later point it may be worth doing a standalone piece of work on the self-doubting conclusion (such as, *I'm a bad-luck magnet*, or *I'm mentally abnormal*) and its linked story, by following the same steps as would apply for doubts in individual situations. The aim is to reorient the person back to the reality of *who he is*—the simple qualities he embodies—as would be evidenced by a video-recording of his behavior in multiple domains of his life, past and present. A further aim is to gain perspective on the imaginal origins of irrelevant fears about who he could be, or could become.

*Case Examples: Invalidating the Doubt**

Below you will see adapted examples of case material given as session summary handouts to clients (further examples in van Niekerk et al., 2014). (In my experience, therapy summaries emailed to clients can be very helpful for therapy consolidation; however, ensure that you follow relevant professional guidelines about secure methods of communication.)

Client 1. (Square brackets designate omission of details for protecting client anonymity; con tricks are labeled in italics.)

Doubt: The books aren't right if they're not facing the same way in the box.

OCD story: It wouldn't look nice (*arbitrary standard*). It would look sloppy and I'd be [bad at my job] (*unsupported link*). It's harder to [do my job well] in messy rooms. I need to run a tight ship. It creates the right impression with other staff. And that's why I need to rearrange the books.

Alternative to the Doubt: The books are fine as they are.

Commonsense view: The function or commonsensical purpose of putting books into a box is (1) so they're protected and (2) so they're easily found. The OCD would like me to make the books look nice as well, but this has little relevance to the commonsensical function. I could tell myself stories of why various noncommonsensical standards for arranging books would make them look nicer, but the fact is I don't. This is why the standard I emphasize is arbitrary and selective. I'd be [better at my job] if I modeled a commonsensical approach… And that's why I don't need to rearrange the books.

Client 2. (Con tricks and therapist comments are in italics.)

Doubt: (*after being served a bottle of water*) The water could be contaminated.

OCD story: The restaurant bottles the water itself—I read this previously. The surface of the bottle wasn't clean; you could see fingerprints on the outside of the bottle. These could have been left by the waitress's hands, which could have been dirty. She's been clearing plates, and people eat from those plates. She could have touched the cutlery, which could have been in people's mouths, and

* Integration of inference-based therapy and cognitive-behavioral therapy for obsessive-compulsive disorder—A case series. *International Journal of Cognitive Therapy*, *7*, 67–82. van Niekerk, J., Brown, G., Aardema, F., & O'Connor, K., 2014.

who knows where they've been. (*Several remote possibilities are brought into the story without any evidence from the senses in the situation.*)

The print could also have been left by a grubby customer who drank from the bottle previously or by the person who bottled the water. The water could have been touched by someone like X, who is not a hygienic person… (*Further remote possibilities are wheeled in.*) That's why I need to ask the waitress for further details to reassure myself that the water and bottle are in fact clean (i.e., that the doubt isn't true). (*But if you asked, that would simply have led to further doubts. In the end, you feel compelled to try to solve a "problem"—the doubt—in reality, but the "problem" has been constructed in the imagination—a difficult task to accomplish as the imagination can potentially sidestep any "solution" in reality. For example, the OCD, via the imagination, could retort that the waitress could be lying when she says that all is okay.*) (Subsequent work allowed formulating the *Alternative to the Doubt* and the *Commonsense View*, which are not described here.)

Relapse Prevention

Review the IBT model and methods and what has been helpful in therapy. Encourage the client to continue applying the insights described above, practicing consolidating with and acting on simple reality, dismissing doubting promptly, and abstaining from residual safety behaviors (e.g., "checking to see if I still have obsessions"). Arrange one or more follow-up sessions to check on maintenance of gains.

CHAPTER 6

Metacognitive Therapy (MCT)

Chapter 1 provided a very brief introduction to MCT for OCD. This innovative therapy was developed by Adrian Wells in the United Kingdom. The discussion below draws on a limited number of texts: primarily the accessible book by Wells (2009), which contains a chapter on OCD, but also the briefer and more generic introduction to the distinctive features of MCT offered by Fisher and Wells (2009). Readers interested in the dividing line between MCT and CBT are referred to the detailed analysis in Fisher (2009). In this chapter I will consider key concepts in the metacognitive model of OCD and the steps in treatment.

Key Concepts

The sections below will introduce you to key concepts derived from the theoretical model that underpins MCT and set the stage for considering their application to OCD.

Metacognition

MCT is rooted in a basic model of psychological disorder called the self-regulatory executive function (S-REF) model (Wells & Matthews, 1996). This offers an account of the cognitive and so-called *metacognitive* (MC) factors implicated in the maintenance of emotional disorders. Metacognition ("meta" means beyond) relates to any knowledge or cognitive processes involved in the interpretation, monitoring, or control of cognition. These can be divided into three types:

- *Metacognitive knowledge* consists of the beliefs and theories that are held about thinking. Positive metacognitive beliefs, such as *Worrying helps me prevent danger,* designate the advantages of engaging in cognitive acts. Negative beliefs, such as *Thinking bad thoughts means that I could be an evil person*, or *I can't control my worry and this could damage my health*, concern negative meanings and consequences attributed to thoughts and cognitive experiences.

- *Metacognitive experiences* relate to the interpretations of and feelings people have about their mental status in a situation, such as déjà vu, worrying about worrying, or assessing the completeness of a memory to establish whether an objectionable act might have been performed.
- *Metacognitive strategies* are aimed at intensifying, suppressing, or changing thinking in order to achieve desirable change. Examples include focusing attention on sources of potential threat so as to be ready to act, suppressing upsetting thoughts, distracting oneself from negative feelings, or using positive thinking to bolster mood. However, in psychological disorder, the strategies used tend to be counterproductive longer term, commonly contributing to a sense of having lost control.

Two Ways of Experiencing Thoughts

According to Wells (2009), humans tend to experience cognitions differently from how they experience perceptions of external events. We tend to be soldered to our thoughts and have little or no distance from them, neglecting to appreciate that they are inner constructions. In MCT this "fused" way of experiencing is termed *object mode*.

This can be contrasted with *metacognitive mode*, during which thoughts and beliefs are viewed as inner events distinct from the self and the world. The individual stands back and observes thoughts as part of the wider backdrop of conscious experience. Experiencing the metacognitive mode requires practice in relating to inner experience in this more detached way, irrespective of the accuracy of the thought. A useful type of metacognitive experience to cultivate as part of MCT is *detached mindfulness*—having an objective awareness of a thought or belief while not engaging in any coping strategy or further conceptual processing and viewing the conscious experience of self as separate from the thought.

Levels of Cognition

Three levels of cognitive processes are distinguished in the S-REF model:

- Automatic processes that run with little or no conscious awareness, which generate intrusions that enter the conscious domain (low-level processing)

- Conscious online (rapid, moment-by-moment, capacity-limited) processing of thoughts and behaviors (labeled *cognitive style*)
- The library of metacognitive knowledge or beliefs stored in long-term memory, which guides online processing

Next, I will consider how the three levels interact with behavioral and emotional responses in the clinical metacognitive model.

Clinical Model

The S-REF explains psychological disorder in terms of *conscious, controlled, top-down* processes and self-regulatory strategies (i.e., those used for attaining emotional or behavioral goals, such as decreasing anxiety or sadness). Instead of using a healthy strategy, which allows distress to resolve, the person uses an unhelpful thinking approach that prolongs, amplifies, and maintains negative emotion. In the S-REF terminology, this unhelpful approach is labeled the *cognitive-attentional syndrome* (or CAS) and—in all psychological disorders—consists of worry and rumination, threat monitoring, and counterproductive thought-control strategies and behavioral coping strategies. The operation of the CAS in OCD, illustrated in figure 6.1, is considered below.

Perseverative Thinking: Rumination, Worry, and Mental Ritualizing

Clients with OCD generally introduce rumination, worry, and mental ritualizing to avoid threat or danger. These perseverative thinking processes are initiated and guided by metacognitive beliefs (discussed below) or rules about thinking and the need to attain certain feeling states. Rumination tends to involve fixation on past events and consideration of "why" questions; the process of worry tends to focus on "what if" questions involving anticipating and coping with future danger.

An example would be of someone with OCD worrying after a car ride that he might have knocked over a cyclist and what this would mean. He would review his memories of the trip extensively to be sure that there were no memory gaps or evidence of a collision. This process would continue until a sense of "sufficient certainty" had been reached. If this were not possible, the person would consider retracing his route in the car, or checking the Internet to ascertain whether a hit-and-run accident had been reported. Two metacognitive beliefs would have motivated the initiation of this train of thought:

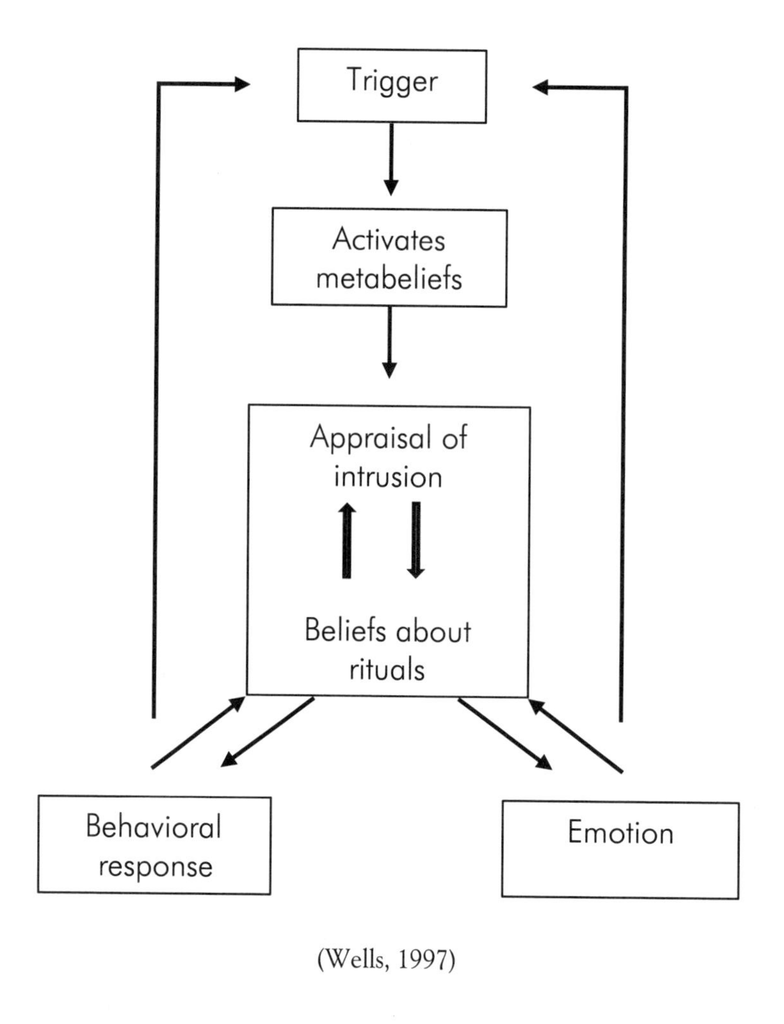

Figure 6.1. The metacognitive model of OCD

1. Positive: It was necessary to establish a sufficient level of certainty for threat to be eliminated.

2. Negative: Thoughts about having knocked someone over have the power, if unattended, to lead to interminable distress and worry.

Threat Monitoring

This represents the coping behavior of using an attentional strategy that involves monitoring potential sources of threat. The person with OCD could monitor a wide spectrum of cues, such as contamination, unacceptable thoughts, and unacceptable features of his personal environment (e.g., lack of tidiness), and also monitor his memory (as in the example described above).

Other Coping Behaviors

Clients can manifest a broad range of behavioral and mental rituals aimed at reducing discomfort and preventing danger. In the metacognitive approach, these rituals are counterproductive for the following reasons:

- They inflate the importance attached to intrusive thoughts through circular reasoning, which leads to the maintenance of this belief.
- They prevent the cultivation of a metacognitive mode of experiencing in which thoughts are considered transient mental events.
- They encourage reliance on inappropriate internal criteria to establish a sense of safety. The application of these criteria tends to activate further intrusions and doubting, as seen in the previous example of the person who was relying on an inappropriate standard for memory (with gaps in recollection considered a basis for concern) and required having a feeling of "sufficient certainty" to be reassured that he had not hit a cyclist while driving.

Metacognitive Beliefs

Previously, a distinction was made between positive and negative MC beliefs (e.g., the benefits or dangers of worry and rumination). A further subdivision is made in the metacognitive model between beliefs about thoughts and beliefs about rituals, worry, and rumination. Furthermore, Wells (2009) describes a special class of negative metacognitive beliefs in OCD called *fusion beliefs*, which concern the powerfulness and importance of negative thoughts (see below).

BELIEFS ABOUT THOUGHTS

Three kinds of fusion beliefs can be distinguished:

- Thought-action fusion (TAF): the belief that intrusive thoughts can cause one to commit unwanted actions (e.g., *Unwanted thoughts of sexually violent acts toward women could cause me to act in a similar way*). This MC conceptualization of TAF borrows from the cognitive therapy use of the term but does not reference a morality dimension (i.e., morality TAF; see chapter 4).
- Thought-event fusion (TEF): the belief that an intrusive thought can cause an external event to occur or means that an event has already occurred (e.g., thinking about a car crash can cause it, thinking about having committed assault means that you did it, thinking that an object

could be dirty makes it so). Nevertheless, bear in mind that *thinking it necessary to assume* an object is dirty to stay safe represents a worry process, which is distinct from TEF.

- Thought-object fusion (TOF): the belief that thoughts and feelings can be transferred into an object, thus making them more real and more dangerous (e.g., thinking that the immoral character of a married person who made a pass at you might have transferred into surrounding items).

BELIEFS ABOUT RITUALS, WORRY, AND RUMINATION

These beliefs constitute a plan for guiding coping responses introduced in response to an intrusion. The plans usually reflect the importance of controlling thoughts and feelings and are aimed at attaining a desired state, which defines the end point of the ritual (e.g., *I can stop washing when I feel clean*, or *I must first tidy and order items in my room so that I can feel in control and be able to study*).

Stop Signals

The MC beliefs about an intrusion determine the view of its significance and the danger inherent in it. The person therefore has to decide when safety or closure has been attained, in the absence of guidance by objective criteria. The OCD strategy for resolving this dilemma is the introduction of and overreliance on inappropriate, subjective criteria that determine the stop signals (also called *stop rules*) for terminating a ritual (e.g., *I can be sure the window is closed after I have checked three times and looked at it from different angles*, or *I have to continue the ritual until it feels just right*).

MCT for OCD

In the MCT model, the obsessive-compulsive sequence is conceptualized as follows:

1. A thought, feeling, or urge (the trigger) activates the person's metacognitive beliefs about its meaning and significance—these beliefs mainly represent fusions between inner experience and the external world (see TAF, TEF, and TOF above).

2. The MC beliefs lead to appraisal of the intrusion as representing threat; the threat appraisal leads to an amplification of anxiety.

3. Intensification of anxiety encourages the intrusion of thoughts and feelings into conscious awareness.

4. Anxiety is also determined by MC beliefs about whether strategies are accessible for neutralizing threat and their expected degree of effectiveness. Such strategies may involve controlling thoughts, feelings, or behavior. The beliefs specify which behaviors or strategies the person introduces and which internal criteria (and stop rules) are applied to define their success or failure. The stop rules are unrealistic and costly to achieve and solidify an attitude incompatible with the person's experiencing thoughts simply as mental events, such as when the "more to be sure" rule leads to ritual escalation culminating in "more" incrementally defining the necessary "minimum."

5. Mental or behavioral rituals are initiated, which are counterproductive in the following ways:

   - Where success of the ritual is measured by monitoring for intrusions, which paradoxically leads to an increase in intrusions
   - Where rituals prevent the person from discovering that beliefs about intrusions and rituals are unfounded—nonoccurrence of the feared event is erroneously attributed to the ritual
   - Where the ritual leads to a widening network of associations with the intrusion, such as where seeing antibacterial wipes triggers the intrusion
   - Where the person repeatedly acts *as if* the intrusion is significant and realistic, therefore reinforcing an object mode of processing, and inhibiting the development of flexibility

Case Illustrations

Below are MCT case conceptualizations of two clients with OCD.

CASE ILLUSTRATION 1

The client is an elderly woman with restricted mobility who experiences obsessions of harming others when she sees people walking past her window, requiring her to look out of the window to reassure herself that she did not harm them. Using the diagram of the metacognitive model of OCD (figure 6.1), the case can be formulated in the following way:

- Trigger: Man has just walked past window → automatic thought: *Maybe I attacked him.*

- Metabelief: *Thinking that I attacked him means I could have done it.*
- Appraisal of intrusion: *That would be an awful thing to do—I would feel terrible.*
- Beliefs about rituals: *If I look out to see if he's okay, I will not feel so anxious.* Stop signal: *I'm safe if I can see the man.*
- Behavioral response: Look out the window
- Emotions: Fear, guilt

CASE ILLUSTRATION 2

The client has contamination obsessions, centered on asbestos, which he copes with by shaking or dusting off his clothing, avoidance of obsession triggers, and washing rituals.

- Trigger: Walking past a shed → automatic thought: *That looks old; there could be asbestos in the roof. There could be fibers in the air. They could be on my clothes.*
- Metabelief: *Thinking I have been contaminated means I have been contaminated.*
- Appraisal of intrusion: *I could get asbestosis.*
- Beliefs about rituals: *If I dust off my clothes and shake my body, I'll get rid of the fibers.* Stop signal: *I'm clean when I feel clean.*
- Behavioral response: Shaking and dusting
- Emotion: Anxiety

Treatment Protocol

The stages of treatment are described in detail in Wells (2009) and are summarized here:

1. Case conceptualization
2. Socialization to MCT
3. Shifting to the MC mode and training in detached mindfulness
4. Modifying MC beliefs about the intrusion

5. Modifying MC beliefs about rituals and stop signals
6. Developing and practicing new plans for processing
7. Relapse prevention

The MCT procedures and techniques described in the sections below fit into Wells's stages of treatment in the following way: case conceptualization (treatment stage 1), client socialization and presenting the treatment rationale (stage 2), detached mindfulness and exposure and response commission (stage 3), exposure and response prevention and modifying specific MC beliefs (stages 4 to 6), and finally, relapse prevention (stage 7).

CASE CONCEPTUALIZATION

Case conceptualization derives from data gathered during clinical interviews and from psychometric assessment.

Clinical interview data. In the first session, ask the client to reflect on a recent episode when he was troubled by obsessions, and ask the following questions to identify components of the model:

- *Triggering intrusion—in MCT this refers to the initial internal event.* Can you remember a recent instance when you noticed an intrusive thought or feeling? What was the thought, feeling, or urge that triggered your emotion or got you ritualizing?
- *Emotion elicited by the intrusion.* When you had the triggering thought or feeling, what emotion did you experience?
- *Appraisal of the intrusion.* What negative meaning did the thought contain about you or about the situation? What is important about the thought? Was having the thought dangerous in some way? What bad things might happen if you continued to have the intrusion? What is the worst that could happen?
- *Metacognitive belief about the intrusion—this involves encapsulating the overall belief about the thought evident in the appraisals.* So, you consider these intrusions as being meaningful and important—more specifically, you believe [*therapist describes* TAF/TEF/TOF]. To what extent do you believe this?
- *The nature of behavioral or mental rituals or neutralizing.* When you thought [*describe the intrusion*], did you do anything to prevent bad things from happening? [*Further probing questions aim to elicit further detail.*] Do you

repeat actions, perform them in special ways, or avoid situations? Did you try to control your thoughts—if yes, what exactly did you do? How often do you engage in the responses and how much of your time do they occupy?

- *Beliefs about rituals and stop signals.* What are the pros and cons of your rituals/coping responses? What is the worst that could happen if you didn't use them? What is the aim of the ritual or coping response and how do you know when it is safe to stop/when you have achieved your goal/how much to do of the ritual?

Psychometric assessment. A number of MCT questionnaires are provided as appendices in Wells (2009), including those listed below; some of these can also be purchased on the Metacognitive Therapy Institute website (http://www.mct-institute.co.uk). Permission is granted for personal use of photocopied versions in client work. Overall, the measures are commendably brief, practical, and clinically relevant. None, other than the OCD-S (see below), has been designed specifically for use in OCD; these are theorized to have broader relevance to psychological disorder. I consider the following measures as being particularly useful in work with OCD to provide assessments of treatment mechanism and symptom severity, at baseline and as therapy progresses.

The Metacognitions Questionnaire (MCQ-30; Wells & Cartwright-Hatton, 2004) is a thirty-item multidimensional measure of metacognition. Five subscales assess:

- Positive beliefs about worry
- Negative beliefs about worry regarding uncontrollability and danger
- Cognitive confidence (i.e., confidence in attention and memory)
- Beliefs about needing to control thoughts and regarding the negative meaning and consequences of not doing so
- Cognitive self-consciousness (the tendency to focus attention on thought processes)

The MCQ-30 is included in the NovoPsych battery for administration on an iPad (see chapter 2).

The Meta-Worry Questionnaire (MWQ; Wells, 2005a) is a seven-item questionnaire that assesses the frequency of metaworry and belief in the dangers of worrying (e.g., going crazy, losing control of worry, making oneself ill with worry).

The Thought Fusion Instrument (TFI; Wells, Gwilliam, & Cartwright-Hatton, 2001) contains fourteen items assessing the level of conviction in various fusion beliefs. Item examples are:

"If I have thoughts about harming myself, I will end up doing it." (TAF)

"My thoughts become reality. If I think something, it will come true." (TEF)

"If I think things are contaminated by other people's experiences, it means they are contaminated." (TOF)

The Obsessive-Compulsive Disorder Scale (OCD-S; Wells, 2009) briefly assesses OCD symptom severity over the prior week and the use of various safety strategies.

CLIENT SOCIALIZATION AND PRESENTING THE TREATMENT RATIONALE

Highlight the common experience of intrusive thoughts and introduce the idea that the client's beliefs about, and how she relates to, her thoughts and feelings cause the OCD. Therefore, changing these beliefs and relating to inner events differently can remove the problem. Highlight the dysfunctional outcomes of various response strategies—they maintain beliefs about the importance of thoughts and therefore perpetuate anxiety. This can be illustrated with in-session behavioral experiments (such as a thought-suppression experiment). Illustrate MCT socialization by making reference to the client's unique OCD presentation.

To illuminate the role of unhelpful beliefs and ritualizing, Wells (2009) suggested asking clients (a) why they continue to have problems if their coping strategies are effective, (b) how many times their checking proved their obsessions to be accurate, and (c) what has gotten in the way of learning that they can ignore their obsessive thoughts.

The treatment rationale is straightforward: the therapist conveys that the client's beliefs about thoughts and about needing to act on them confer unjustified importance to the thoughts, which have been fused with reality. The aim of treatment is to view them and respond to them as being inconsequential, transitory mental events.

DETACHED MINDFULNESS

Wells (2005b, p. 340) describes detached mindfulness (DM) as "a type of inner-awareness, but in the absence of effortful processing of the self." The components are:

- Meta-awareness (being aware of your thoughts)
- Cognitive decentering (viewing thoughts as events, not facts)
- Attentional detachment (recognition that attentional focus is flexible and can be shifted)
- Low level of conceptual processing (reduced inner dialogue)
- Low goal-focused coping (goals of threat-avoidance or -removal are not a priority)

Training in DM in OCD is focused on cultivating a more adaptive way of relating to obsessive thoughts and strengthening the metacognitive mode of processing. It is considered a prerequisite for proceeding to modification of metacognitive beliefs (i.e., TAF, TEF, TOF).

The first step involves reviewing several instances where OCD is triggered and assisting the client in being able to identify the initial triggering intrusive thoughts or doubts.

Next, help the client to respond differently by using DM. One approach is to use a free association task in which you read a list of nouns, asking the client just to observe her thoughts in a detached way as the passage of events in the mind, without trying to influence or control them in any way. The task may need to be repeated and modified a number of times. (Other DM techniques are provided in Wells, 2005b, and Wells, 2009.)

When success has been attained, the next step is to instruct the client to observe her mental contents in a similarly detached way, allowing her thoughts to roam freely and any thought, including obsessions, to enter her mind.

Hereafter the aim is to inculcate greater mindfulness of the dividing line between the sense of self and the intrusive thought. In the first condition, the person holds a neutral thought in mind (e.g., of an apple). Request that the client take a mental step away from the thought and reflect on the divergence of her observing self from the mental event of having the thought. In the second condition, the person attempts the same with an obsessive thought. Wells (2009) recommends that in clients with OCD, DM be practiced over several sessions, but this is not intended to be a task that clients practice in response to obsessions. Rather it is a learning exercise that helps to develop new ways of relating to intrusions generally.

EXPOSURE AND RESPONSE COMMISSION

Exposure and response commission (ERC) is a further MCT technique used for enabling greater meta-level experiencing of the intrusion, in order to develop the ability to engage in detached mindfulness. This involves the client being instructed to *hold the intrusion in mind* while performing rituals, thus changing the normal contingency of rituals being aimed at removing the intrusion. The client can do this by repeatedly subvocalizing a verbal thought or visualizing an image while performing the ritual. Introduce the exercise as a strategy to assist the client in attaining distance from thoughts and discovering that they can be unimportant. For example, a person who ritualistically cleans an object that has been morally polluted might repeatedly tell himself that the object was in fact morally contaminated, while in the process of cleaning.

EXPOSURE AND RESPONSE PREVENTION IN MCT

ERP in MCT follows a similar procedure to its use in behavior and cognitive therapy (chapters 3 and 4), but it is also different in important respects. It can be brief and practiced only a few times as its target is specifically a reduction of the cognitive-attentional syndrome (CAS)—the pathological pattern of cognition that consists of unhelpful attentional strategies, perseverative thinking, and other counterproductive coping responses (see "Clinical Model" above). Another ERP target is the testing of metacognitive beliefs (see following section). Carefully assess all overt and covert coping responses; this analysis forms the basis of an instruction for the client to eliminate or delay all the dysfunctional coping behaviors—conceptual processing (worry or rumination), any threat monitoring, and ritualizing (ritualizing can be delayed to a ten-minute period at the end of the day, but this is not compulsory). This allows predictions to be tested about the accuracy of metacognitive beliefs and the feared consequences of reducing or omitting coping responses.

MODIFYING SPECIFIC MC BELIEFS

The initial focus of treatment is on facilitating the experiencing of obsessional thoughts and feelings in the metacognitive mode, through the training of detached mindfulness (see above). Hereafter the focus shifts to the restructuring of metacognitive appraisals and beliefs. Both verbal and behavioral techniques can be used.

Verbal techniques. Wells (2009) stresses the importance of working with the client to articulate fusion beliefs about obsessions, which are often implicit. Common generic belief content is that intrusions are important or powerful in some way; a situation-specific formulation benefits from incorporating the details relevant to the specific intrusion.

Standard verbal strategies for restructuring fusion beliefs include questioning the supporting evidence and seeking counterevidence, including whether there is a plausible mechanism to account for a fusion effect:

- Do all of your thoughts have the power to cause events, or only some?
- How many times have you checked again and found the iron was still switched on? What does this tell you about how important the obsessive thought really is?
- Did you have any similar thoughts before your OCD started that you did not neutralize? What happened?
- How does it work that a thought can enter and remain in [whatever the chosen inanimate object is] and contaminate it in that way?
- Is a thought about an event the same thing as a memory of the event?

Maximize the therapeutic leverage afforded by *cognitive dissonance* (a situation in which conflicting, simultaneously held beliefs cause discomfort) by exposing contradictory beliefs. For example, with a client who has obsessive thoughts that he might have offended friends, reflect on the fact that he knows his friends to be the kind of people who would have said something. Or a person who believes he might have attacked someone can consider the kind of person he knows himself to be.

Behavioral experiments. Appropriate behavioral experiments for likelihood TAF (morality TAF is not part of MCT) have been described in chapter 4 and will not be repeated here. However, Wells (2009) offers interesting suggestions in respect of the other categories of fusion.

TEF can be made to apply to future events (*If I think a loved one will come to harm, it may happen*), or past events (*If I think the floor has poop on it, it must be dirty*). Prospective TEF can first be addressed by testing the ability to cause positive events via modification of thinking (see chapter 4), as part of a graded strategy. Thereafter the ability to cause negative events can be tested. Again, a graded strategy can be used to enhance compliance (e.g., starting with thinking about causing minor harm to pets and progressing to causing harm to loved ones).

The above is more straightforward when the time frame of the predicted catastrophe allows direct testing. Where the time frame includes the distant future, consider the possibility that this represents the use of an avoidant strategy. One option is to question the basis for the prediction of a longer, rather than a shorter, time frame, and test for a shorter time frame anyway ("Even though you say that this [disaster] can happen at any point in your life, let's find out what happens over the next two weeks"). Or, the obsessional content can explicitly be modified to reflect a briefer time frame ("Let's see if we can cause your guinea pig's fur to fall out by Sunday").

Retrospective TEF can be addressed by having the client perform *adaptive checking* rather than the usual compulsive checking aimed at assuaging worry and anxiety. Here the checking is explicitly designed for and aimed at *establishing the veracity of the thought*, thus allowing data to be generated to support a view of the intrusion simply being an irrelevant thought. In this way, metabeliefs about the importance of thoughts can be tested.

TOF can be tackled by testing the client's ability to identify a thought-contaminated item. For instance, you can ask the client to contaminate one card by thinking impure thoughts, and then mark that card. Afterward, ask the client to close her eyes and pick the contaminated card from a selection of cards. If the client cannot identify the contaminated item, this can be considered in light of what has been revealed about the nature of the contamination—ask the client, "If the contamination is entirely subjective (only thoughts and feelings), then what is the appropriate response?"

Other experiments address the predicted consequences of object contamination, either in terms of discomfort or distress, or real events. An example of the former is where multiple objects are intentionally contaminated, leading to fading of the thoughts and feelings, rather than the predicted opposite. An example of the latter is where a client worries that he will start acting in immoral ways if he sits on a public bench in an area frequented by drug dealers. First, the feared predicted immoral behaviors are defined in a way that makes it possible for them to be clearly observed and measured; thereafter the client sits on the bench and then records whether any immoral actions follow (e.g., sitting on the same bench as somebody who litters or smokes, to test whether he starts littering or smoking). A graded strategy may be required if the client's anxiety about the exposure is very high.

Wells (2009) discusses the use of the therapy prop of a water spray can to aid in the treatment of contamination fears. Contaminated items can be touched with a piece of paper, which is dipped into the water in the spray can. By spraying the "contaminated" water, various beliefs can be tested, such as whether obsessive thoughts and feelings endure, or whether the problem is one of real danger or of thoughts about danger.

Modifying beliefs about rituals. Wells (2009) generally suggests that this module follow work on the metabeliefs about the intrusions, but it may also be necessary when such work requires ritual prevention experiments. Informed by the case conceptualization, the therapist engages the client in a Socratic dialogue to broaden the client's awareness of all the unhelpful consequences of ritualizing. The following questions allow identification of the pitfalls of ritualizing:

- Can you think of any problems associated with using rituals to cope?
- Have they helped you make progress in overcoming your OCD? How may they hold you back?
- Are there negative effects on your ability to cope, your well-being, your environment, or people in your life?
- Do the rituals help you in the long run? Why not?
- Are there ways in which your ritualizing might prevent you from being able to see things clearly or realistically?
- Do they stop you from putting your fears to the test?

Wells (2009) also recommends explicitly identifying and negatively reframing the perceived advantages of ritualizing. The following are examples of questions that can be used and the negative reframe which can be applied:

- Do you see advantages of performing your rituals? (Reframe: Is this also a longer-term solution? In what ways may they perpetuate the problem?)
- Is there anything positive about your rituals? (Reframe: Is there a better way of achieving this?)
- Do your rituals help you to prevent harm, or loss of control of your mind? (Reframe: May they stop you from knowing if the harm is real or imagined? Was there a time when you had the obsession and could not perform a ritual? What happened?)

Client predictions about the harmful consequences of ritual abstention include catastrophic events, unrelenting worry or anxiety, and impairment in functioning, such as not being able to concentrate. These beliefs, and other beliefs about the need for rituals, can be explicitly tested in behavioral experiments using an ERP format (see previous section).

Modifying attentional strategies and criteria for knowing. MCT recognizes that clients sometimes adopt dysfunctional attentional strategies as part of a pathology-maintaining processing plan. For example, as a result of her fusion beliefs, a person with contamination obsessions may first scan any object she touches for evidence of red spots ("which could be HIV-infected blood"). This is a strategy for staying safe; however, it will increase the sense of lurking danger owing to enhanced detection of red spots and lead to a stronger need to ritualize. As part of a comprehensive strategy, the counterproductive attentional prioritizing will also have to be addressed—in this example, the person will have to cease her prescanning of objects she touches.

Further, MCT addresses obstructive criteria for knowing that a task has been satisfactorily accomplished, in other words, criteria that define the end point or stop signals for rituals. Examples include excessive force applied when turning the door handle to accept that it is locked, having to attain a "feeling" of one's body or an item being clean in the absence of objective information affirming the opposite, or having to have a clear, detailed recollection of how time was spent to accept that an event did not take place, rather than just the absence of a memory of the event.

Addressing unhelpful criteria and attentional strategies requires that a new processing plan be developed with the client and practiced, as in the example below:

OCD Plan:

- I should look as I press the lock button on the car key
- Listen for the click
- Pull the door handle as hard as I can, hold
- Let go, look at the door and think *locked*
- Walk away from the car, think of the door being locked

New Plan:

- Look at pressing the lock button
- Listen for the click
- Do not pull the door handle, do not look at the door and think *locked*
- Walk away
- Apply detached mindfulness to intrusive doubting thoughts and tolerate discomfort; remember that intrusive thoughts are just that, not facts

RELAPSE PREVENTION

Wells (2009) recommends that the final two sessions of therapy be devoted to developing a therapy blueprint, which provides a tailored summary of the following:

- Case formulation
- The evidence challenging specific metacognitive beliefs about intrusions
- The disadvantages of performing specific rituals
- The old and new plans for processing and behavior

This will benefit relapse prevention by summarizing what was helpful in therapy, which can serve as a written resource to guide the future management of any residual symptoms or the response to challenges that have the potential for reactivating symptoms.

CHAPTER 7

Acceptance and Commitment Therapy (ACT)

ACT (referred to as a single word, *act*) was developed by University of Nevada psychologist Steven Hayes and his collaborators. ACT has injected fresh ideas into the CBT arena, which have stirred up spirited debate (see Gaudiano, 2011), and over recent years has attracted considerable attention and gained in popularity. However, despite encouraging research findings to date, as the new kid in the CBT neighborhood, ACT needs to establish whether it can convincingly add to the efficacy of older approaches.

ACT was developed to target basic transdiagnostic processes underpinning most or all psychopathology (see below). Nevertheless, clinical psychologist Michael Twohig and colleagues recently developed an ACT protocol for OCD and tested it in preliminary research studies (see Twohig, 2009; Twohig et al., 2010). First, I will briefly consider ACT's philosophical and theoretical background in functional contextualism and relational frame theory (RFT). Second, I will discuss the key psychological processes identified in ACT as being instrumental in psychological disorder. Third, ACT for OCD will be considered in more detail. The references cited in this paragraph and Russel Harris's (2009) excellent introduction to ACT are the key source texts for this chapter. They draw on the pioneering work by Hayes and colleagues (see Hayes, Strosahl, & Wilson, 2012).

Key Concepts

As Louis D. Brandeis famously said, "If you would only recognize that life is hard, things would be so much easier for you." The ubiquity of suffering that he speaks of is a foundational principle in ACT, as described below.

Suffering, Mind, and Language

ACT is distinguished by a view of significant distress as being a *normal* part of the human experience, supported by studies suggesting that marked distress tantamount to a diagnosis of psychological disorder is in fact experienced by many over the course of their lifetime (e.g., Moffit et al., 2010).

An ability that distinguishes humans from nonhuman animals is human language: the set of symbols we use that allows us to convey ideas ranging from the practical and mundane to very high levels of abstraction and complexity, even allowing us to create transcendent beauty in poetry and literature.

Public language and *private language* can be distinguished. Examples of the former are talking, singing, painting, and gesturing; the latter includes thinking, analyzing, visualizing, imagining, remembering, worrying, fantasizing, and ruminating. In psychological discourse, private language is commonly described as *cognition*.

Language abilities make it possible to evaluate almost any aspect of human experience negatively ("Everything seems beige," as a person with clinical depression described it to me). This exponentially increases the potential for human suffering.

In ACT, the concept of "mind" is a metaphor for language and is routinely used in therapy.

"Third-Wave" Therapies, Relational Frame Theory, and Functional Contextualism

Three waves of behavioral therapies emerged over the last century. The "first wave" consisted of approaches developed largely from the 1940s through the 1960s, in which psychopathology was conceptualized in terms of observable and measurable problem behavior, acquired through classical or operant conditioning. The aim of therapy was unlearning problem behaviors and substituting these with more adaptive behaviors; little importance was attached to thoughts and feelings.

In the 1950s and 1960s, the "second wave" of behaviorism took off, spurred on by the evident limitations of the first-wave approaches. (The philosopher Sidney Morgenbesser, perhaps a bit unfairly, criticized a leading proponent of first-wave behaviorism, B. F. Skinner: "Are you telling me it's wrong to anthropomorphize people?"; quoted by Dennett, 2014). Here, the view was that potentially modifiable, irrational, and distorted thoughts produced psychopathology, and CBT sought to substitute these with more accurate, helpful, distress-alleviating alternatives; the aim is symptom alleviation through cognitive change.

Striving to improve upon the second wave, starting in the 1980s, Steven Hayes and colleagues developed relational frame theory, to extend first-wave behavioral principles to account for the effects of language (cognition) on human behavior (remember, cognition was ignored in the first wave).

In a nutshell (RFT is fairly dense theoretical reading), RFT considers language to be operant *behavior*: what we say, or think, depends on past reinforcement (being rewarded for saying or thinking those things in similar circumstances). This would not deviate from first-wave behaviorism, but an additional element is added when RFT states that language is also a *relational operant*: language use involves relating stimuli in a way that changes how we react to those stimuli (e.g., items, people, thoughts, feelings); in other words, it changes the function of the stimuli. This process, called relational framing, interacts with and often overrides conventional conditioning processes. It provides a method whereby arbitrary stimuli can acquire an anxiety-provoking function in the absence of the person having any actual experience of them.

Consider this example: My *friend had a breakdown because of anxiety; therefore, anxiety is dangerous. I'm anxious; therefore, I must be in danger; therefore, I should act to reduce my anxiety*. Here, relational framing resulted in the feeling of anxiety acquiring the function of stimuli with a property of being dangerous. The pitfall is that language may end up determining behavior in a way that does not reflect the reality of the situation (the feeling of anxiety is not dangerous!) and this may be difficult to unlearn.

How does RFT inform ACT? The developers of ACT place considerable emphasis on the therapy being carefully derived from a well-substantiated underlying theory. In ACT, inner experiences are considered as being responses to environmental events rather than independent causes of behavior. Therefore, a better way to address problematic thoughts and feelings may be changing the *context* in which they are experienced, rather than trying to change their *content or frequency*, which is the focus of conventional CBT. Modification of this context, through acceptance and mindfulness strategies (see below), becomes the route for changing an unhelpful function (or consequence) of thoughts in terms of an unwanted behavioral consequence. But the focus is still on eliminating problem *behavior*, similar to first-wave behavior therapy.

It is this focus on modifying the context (or how one relates to one's thinking), rather than the content, of distressing thoughts and feelings that is a defining characteristic of third-wave behavior therapy. Other therapies under this heading include dialectical behavior therapy (Linehan, 1993) and mindfulness-based cognitive therapy for depression (Segal, Williams, & Teasdale, 2013). These represent a transformation and integration of earlier phases, rather than a discarding or replacement (Twohig, 2012).

Finally, the philosophical framework that underpins ACT and RFT is called *functional contextualism* (FC). This implies a pragmatic and nonideological stance where the goal of theory is to successfully predict and influence behavior, rather than to establish the "true" essence and nature of the relevant psychological processes. Instead, ideas are true only in as far as they allow accurate prediction and influencing. Further, similar behaviors can have different functions depending on the context. And the same applies to inner experiences such as thoughts and feelings; none is inherently problematic, negative, or positive—this depends on how they function for that person (Twohig, 2012). In ACT, the desired function of behavior (its "workability") is that which helps the client live a more vigorous, meaningful, values-consistent life (see Hayes, Levin, Plumb-Vilardaga, Villatte, & Pistorello, 2013).

Clinical Model

ACT considers that the mind can be put to good or sinister use. For example, it can free people from the requirement of having to experience a situation to know if it is safe or dangerous, but it can also lock people into a harmful behavioral pattern, going against their values and not to their longer-term benefit, thus diminishing vitality, narrowing horizons, and prolonging pain.

Six interconnected psychological processes result in people "getting stuck" in psychopathology (their functionality is judged by the context of their use—in some situations they may be helpful, but in others they exact a psychological cost). These are described below, including how they are addressed in treatment to promote greater *psychological flexibility*—the ultimate goal of ACT (see figure 7.1). Psychological flexibility is defined as "being able to contact the moment as a conscious human being more fully as it is, not as what it says it is, and based on what the situation affords, persisting or changing in behavior in the service of chosen values" (Hayes et al., 2013, p. 8). Or more simply: "the ability to be present, open up, and do what matters!" (Harris, 2009).

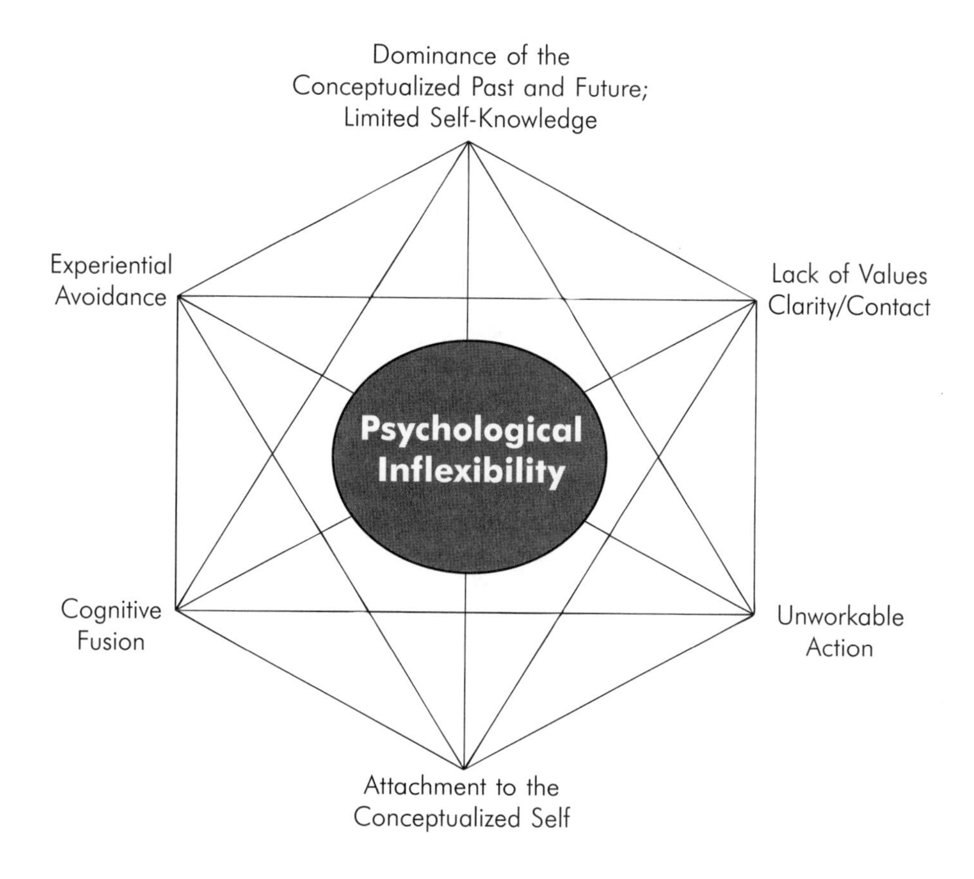

(Adapted from figure 2.1 in Russ Harris, ACT Made Simple, 2009)

Figure 7.1. ACT model of psychopathology

Cognitive Fusion

Cognitive fusion happens when people take their thoughts literally, becoming inseparably entangled, without being aware of the process of thinking itself, such as when getting caught up in a repressive rule about what should be important in life. The consequence may be that one's thoughts end up having excessive influence on one's actions. In ACT, this is described as "being pushed around/ordered around by your thoughts" or "getting hooked/entangled in/carried away by your thinking."

The potentially destructive impact of fusion with evaluative language is forcefully conveyed in the film *Bird Man*, where an incensed playwright acted by Michael Keaton sneers at a *New York Times* theater critic who had just callously informed him that she was going to kill his play simply because she hated him and everyone he represented: "You just label everything!... Do you know what this is (pointing at a flower)?... You don't; you can't see this thing if you don't have a label! You mistake all those little noises in your head for true knowledge."

Defusion is a key strategy and involves learning to mindfully observe thinking as it happens, stepping back and not getting entangled, watching thoughts come and go as if standing on a balcony looking down on pedestrians while not interfering in the direction they take.

How does one attain this? ACT has a rich cache of techniques to draw on, including metaphors, analogies, and experiential exercises. The following is a sample of defusion techniques (some of these and other techniques will be described in more detail below):

- Repeating thoughts until only sounds remain
- Labeling thoughts (*I am thinking that...*; *I am having a feeling of...*)
- Thanking the mind for a thought
- "Taking your mind for a walk"—the therapist walks behind the client while talking to represent the chattering mind
- Writing thoughts on cards
- Acting in a way that directly contradicts a thought
- Attaching thoughts to clouds and letting them drift by (an imagery mindfulness exercise)

Experiential Avoidance

Experiential avoidance is "when a person is unwilling to remain in contact with particular private experiences (e.g., bodily sensations, emotions, thoughts, memories, behavioral predispositions) and takes steps to alter the form or frequency of these events and the contexts that occasion them" (Hayes, Wilson, Gifford, Follette, & Strosahl, 1996, p. 1154). The person tries to eliminate, escape, suppress, or avoid private experiences.

Why do humans do this? It's because we're constantly using problem solving to address problems successfully in the external world (poor soil in your garden—add compost; antisocial neighbors—complain, call the police). And because this approach works so well with events out there, it's very tempting to overuse it in addressing "problems" in the inner world—if a hammer has worked so well for you, everything looks like a nail.

However, there's a myriad of unintended negative consequences. For example, let's say a heterosexual person copes with an obsession about potentially being gay by spending lots of time trying "to figure it out," sometimes testing his sexual responses, or avoiding obsession triggers, such as looking away from attractive men on the street. These are some of the costs:

- He spends a huge amount of energy and mental effort in the battle against anxiety and his own mind—there may be short-term benefit, but longer term, the problem remains.
- His battle against anxiety might irritate his partner because he frequently seems to be preoccupied and mentally absent, thus compromising his valued roles.
- He may reduce his engagement with interesting and life-enriching activities.
- He may become reliant on alcohol or medications to "take the edge off" or "just take a break from it all."
- He may grit his teeth and engage with feared situations anyway.

The last option is likely a less unhelpful behavior than the ones listed before, but in ACT this would still be a form of experiential avoidance as the person is still struggling with his feelings, wishing them away—this would be *tolerance*, not true acceptance. In fact, research studies have found experiential avoidance to be instrumental in a wide range of pathologies (see Hayes et al., 1996).

The antidote to experiential avoidance is *acceptance*, which in ACT is not of the "grim resolve" variety. Instead, acceptance can be better described as a *willingness* to experience the distressing experiences and emotions encountered over the course of value-consistent action. Flaxman, Blackledge, and Bond (2011) describe the following experiential exercise to convey the essence of the concept: hold a block of ice in your hand and focus on the sensation; notice thoughts and feelings come up; when hooked into a thought, bring your attention back to the sensation in your hand. The (objectively unpleasant) sensation is simply attended to over the course of valued action (learning about ACT) for what it is, rather than what our mind (thoughts) tells us it is.

Mindful attending to unpleasant experience may make such an experience less aversive; however, ACT stresses that *the goal of mindful attending is to enable value-consistent action, rather than symptom reduction*. If the latter applied, there would always be the chance of residual levels of negative emotion getting in the way of valued living.

Loss of Flexible Contact with the Present

Cognitive fusion and experiential avoidance readily translate into losing touch with our here-and-now experience. Thoughts about the future (e.g., worries, fantasies, plans), or thoughts about the past (ruminations, regrets, nostalgic moping) easily pull us away from the moment. However, even though this is not in itself a negative, there may be costs associated with a pronounced asymmetry in favor of the mind spending long periods disconnected from the moment or with an inability to flexibly attend to the present moment (or "contact" present experience, in ACT-speak).

Preoccupation with the past or future, or even being absorbed in evaluative thinking about the present situation, easily detracts from learning from and acting in an effective way in the present. This can hold back the development of self-knowledge. Fully, mindfully attending to the present moment is difficult; however, even inconsistent contact, particularly when repeatedly getting hooked in by distressing thoughts, can facilitate more effective learning and action. Mindfulness homework in ACT is used to practice a mode of mind that is more curious, flexible, and appreciative, and less fixated on problem solving and judgmental evaluation.

Attachment to a Conceptualized Self

ACT distinguished *self-as-content* from *self-as-context*. Self-as-content has also been labeled the "described self" and designates our evaluative stories about who we are; frequently experienced or salient thoughts, feelings, and memories are viewed as self-defining experiences. Fusion with either an overly negative (e.g., depression) or positive (narcissism, grandiosity) conceptualized self can be problematic.

Instead, ACT seeks to promote stepping onto the observer's platform of the self-as-context (also called the observing/noticing/transcendent self). This is designated by the use of "I" in language (e.g., I did, I felt) and refers to the perspective from which observations about the experience of the self are made. When consciousness is recognized as being distinct from and not defined by the content of one's self-description, distressing content can potentially become less threatening and less disabling.

Examples of techniques used to facilitate accessing the self-as-context (more detail below) include mindfulness exercises, language reformatting, metaphors, and experiential exercises, such as writing a letter of advice to the person struggling now, from the vantage point of being older and wiser.

Values Confusion

Values define the positive qualities people wish to embody in their behavior over the course of their lives. In ACT, values are approached as process variables, that is, not attained by a certain date (which is a goal), but rather as ongoing signposts for behavior that imbue a person's life with an increased sense of vitality, meaning, and purpose.

A linked concept, put forward to emphasize the role of ongoing behavior, is *valued living*. This can be thought of as ways of acting that give access to relatively stable, long-term sources of positive reinforcement (Flaxman et al., 2011).

ACT therapists stress that the values identified should be personally meaningful to the client. Behavior can be motivated by "want to" (or, in ACT jargon, under *appetitive* control—appetitive behavior increases the likelihood that a need will be satisfied). Research suggests that "want to" behavior avoids the disadvantages of motivation by "must," "should," or "have to" (which represents being under *aversive* control—"the tyranny of the musts"). The latter can reflect mere compliance with socially prescribed values (others telling you how to be) or fusion with a personal rule to avoid guilt or shame (*To be a good person, I always have to put in maximum effort*).

Therapy can assist in value clarification and identification of coordinate behaviors that define the value. Defusion strategies can help in clarifying values and maintaining them as "want tos" rather than "have tos"—as moment-by-moment free choice rather than duty-bound obligation. This will result in valued living being maintained through intrinsic reward, despite valued goals sometimes not being attained (such as when, despite the client's acting supportively to his friend, the friend's decision making results in the breakdown of the friendship).

Unworkable Behavior

As a form of behavior therapy, ACT almost always involves pursuing behavior-change goals. Unworkable behavior refers to behavioral patterns that detract from valued living, such as inaction, procrastination, impulsivity, or avoidant coping. For example, a person feeling depressed about her OCD may retreat from her social life, spending a lot of time sleeping, watching TV, and eating junk food.

Instead, ACT encourages making a commitment to and implementing value-guided action. In the sometimes-complex behavioral configuration in any situation, the aim is to increase the proportion of valued behavior and decrease the proportion of unworkable behavior (such as if I value working efficiently and expanding my knowledge, I would continue studying for my exam but turn off Spotify and my cellphone, which are proving distracting).

It may be helpful to plan and schedule valued behaviors, or clarify which valued behaviors to introduce when opportunity arises. Commitment is approached as an ongoing momentary free choice rather than a monolithic life goal, to avoid clients' feeling overwhelmed. Any established behavioral technique, in the service of ACT goals and consistent with its theory, can be used, such as skills training, exposure, stimulus control, and shaping.

ACT for OCD

All of the processes described above feature across disorders. However, the content and frequency of problem behavior (which includes language) varies between different disorders. Therefore, work has been done to adapt ACT for OCD. A summary of the treatment manual developed by Twohig and colleagues (see Twohig, 2009; Twohig et al., 2010), who adapted the ACT manual by Steven Hayes and colleagues (Hayes et al., 1999; Hayes, Strosahl, & Wilson, 1999), is provided below. For more detail, the reader will benefit from reading the listed publications; Twohig's (2004) abbreviated treatment manual is available on the Association for Contextual Behavioral Science website (which provides a wealth of ACT materials and resources).

OCD Model and Case Conceptualization

Human language abilities produce two beliefs that underpin OCD pathology: (1) negatively evaluated inner experiences pose a threat, and (2) these need to be diminished or controlled. Therefore, cognitive fusion and experiential avoidance (see above) are center stage. However, attempts to suppress obsessions are not effective, and attempts at regulating inner experience miscarry and sabotage quality of life.

The case conceptualization considers the client's OCD problem and history of the problem in terms of the six aforementioned processes (see "Clinical Model"). There will be evidence of fusion with obsessions leading to pronounced experiential avoidance and unworkable behavior (ritualizing, avoidance, and other control strategies). Fusion with a conceptualized self is perhaps most evident where obsessions relate directly to the self (e.g., *I could be dangerous/abnormal*). Inevitably, control strategies encroach on valued living.

Therapist Stance

ACT endorses a humanitarian, nonhierarchical therapist stance toward the client. The client is not damaged and not in need of being repaired. The therapist likely has not had OCD, but both therapist and client have struggled with psychological issues, endured life's travails, and experienced setbacks. (This explicit articulation of an egalitarian value—not exclusive to ACT—is an important safeguard against an unhelpful definition of "expertise" in a psychological therapy setting: hierarchical, patronizing, disempowering of the client, potentially self-serving, and, therefore, antitherapeutic). The therapist's job is to help people get unstuck and move forward, and the client is the expert on his or her difficulties.

Session Structure

Sessions follow a very similar structure to what is typical in CBT:

- Assessment of functioning (OCD symptoms over prior week)
- Reactions to last session
- Review of homework
- Session topics
- Assignment of homework

Treatment Protocol

Twohig's (2004) manual is designed to be administered over eight one-hour weekly therapy sessions. He advises that although a sequence of steps is described, a flexible approach is called for in the sequencing of therapy, depending on which psychological processes are at the fore in the client presentation. Rather than offer a session-by-session account, I will describe key issues and techniques that characterize the main phases of therapy as outlined by Twohig.

ASSESSMENT

The initial assessment described in chapter 2 also applies when doing ACT. However, as part of ongoing assessment over therapy (see below), the focus is on how the information obtained supports a view of the client's problem in terms of ACT processes.

THERAPIST INTRODUCTION OF THE MODEL

You may introduce the model in the following way:

There are two alternatives. Some therapists will focus on changing how you think and feel. That may be an option. An alternative, instead of trying to help you win the struggle you've been in, is to help you step out of it. This focuses on the things you do that have kept you struggling and seeks to change them. It's intensive work, focusing on the relationship you have with your psychological experiences, thoughts, feelings, and so on. It will make better sense to you as you experience it over the course of therapy.

If the client consents to ACT, advise him that therapy can stir up a range of emotions, to be prepared for this, and to suspend final judgment until he has given therapy a fair chance.

THE FUTILITY OF THE WRONG KIND OF CONTROL

The client needs to have a basic understanding of OCD, namely the difference between the obsession and the control strategies, or *control agenda* (behavioral or mental rituals, including "figuring it out," neutralizing, thought suppression, avoidance of obsession triggers, and using medications, alcohol, or illicit drugs to cope). At this early point, he is unlikely to appreciate that the obsession can occur without necessarily being followed by a compulsion.

Typically, the initial focus in therapy is on comprehensively examining the effectiveness of the client's control strategies. The control agenda usually facilitates some short-term relief, which should be acknowledged, but the main focus is on bringing to light that struggling against the obsessions has made things worse—it has not offered a long-term solution, *and it never will.* Commonly, clients readily acknowledge the former, but they tend to entertain lingering hopes of discovering a technique or an insight (in therapy) that works, or that they will finally "figure it out" and achieve the peace and certitude of not having obsessions and anxiety.

Normalize the client's attempts to rid herself of distressing inner events ("We all do this sometimes") and collaborate with her on trying to figure out what "works" to decrease the obsession. Ultimately, even though the client's mind tells her there is a way out, her experience tells her that none of the OCD strategies works. This process facilitates the development of a client's thinking to reach a stage of *creative hopelessness*—where she is open to considering alternatives, such as "stepping out of the battle."

The strategies below help the client examine the effectiveness of her control strategies.

Metaphor. ACT relies heavily on metaphor and analogy to communicate concepts. The above can be illustrated using the "digging yourself out of a hole" metaphor:

In the metaphor, the person is placed in a field, blindfolded, with a small bag of tools. She is told her job is to run around the field blindfolded (which represents getting on with her life). But sooner or later she falls into a hole (which represents the obsession). She opens the bag of tools and finds a shovel. She starts digging to get out, this way and that way (methods used to reduce the obsession). The hole gets bigger (the OCD and its costs increase). Yet, no matter how motivated she is, or which methods she uses, there is no way out. But she keeps on digging. She's asking the therapist for a mechanical excavator!

The aim of the metaphor is to convey the merit of decreasing the focus on reducing obsessions and understanding the difficulty inherent in controlling them. The client needs to act with courage to let go of the shovel and think outside of the box—consider a strategy other than digging. Because of the power of old habits, she will have to take this courageous action many times.

Homework. To further the above goals, assign homework in which you ask the client to list the costs of OCD and record all the control strategies ("your ways of digging"), judging how useful they have been in solving the problem.

THE PARADOX OF CONTROL

The first step in therapy is to subvert the client's control agenda; the next is to expose the paradoxical effects of control and to continue to challenge the rule that the obsession shouldn't be experienced and that the rituals present a solution—strengthening a view that *control is the monster and its henchmen are escape and avoidance.*

Introduce the idea that the use of language has enabled humans to have great success in manipulating and modifying the external environment ("If you don't like it, figure out how to get rid of it and then get rid of it")—deliberate, purposeful, conscious control works very well here. However, the rule works less well for controlling mental experiences, such as the thoughts and feelings automatically introduced into the situation by one's history. Here the rule tends to be "If you aren't willing to have them and try to eliminate them, you'll get stuck with them." Even though the control strategies may offer short-term OCD relief, in the longer term they backfire. However, "being willing to have it *to get rid of it*" is also flagged as being a problem because it signifies nonacceptance. This introduces the important concept of willingness to have uncomfortable experiences as a step toward the ACT goal of acceptance.

The concept of willingness is nicely illustrated by using the "child in a supermarket" metaphor. In this metaphor, ask the client what is likely to happen when a parent and child walk past the sweets or toys section in a store. Then, equate the client's attempts to control the obsession with inadvertently training children how loudly they should scream in order for the parent to buy them candy. The hope is that the client will start appreciating that a program of control rooted in nonacceptance actively makes the situation worse, and that an attitude of willingness to have a negative experience represents an alternative.

ACT employs numerous experiential exercises to help clients internalize the ACT processes, such as the one below.

The Polygraph Machine

Tell the client that he's hooked up to a polygraph machine and that his only task is to stay relaxed. The machine is ultra-sensitive for detecting anxiety. An incentive to comply with the task is provided by a gun being held to his head—if he experiences anxiety, BAM! The difficulty of emotional control is self-evident and can be contrasted with how much easier it is to control behavior, for example if the task was only to sweep the floor. Next, explain that the human nervous system is represented by the polygraph machine and diminished self-worth and quality of life are represented by the gun (i.e., the self-inflicted costs associated with having the prohibited experience). The metaphor aims to convey the difficulties inherent in successfully controlling thoughts and feelings compared to controlling behavior.

The client's repertoire of OCD control strategies may include substituting negative with positive feelings, on the assumption that positive feelings are easier to control. You may offer a hypothetical reward for the client being able to fall completely in love with an arbitrary person. Would he be able to do it? Thought-suppression exercises (described in chapter 4) for positive, neutral, or negative stimuli can also highlight the difficulty in controlling thinking. These additional exercises expand the client's sense of the range of private events that are not amenable to easy mental control—getting rid of anxiety or negative thoughts, creating positive emotions, and so on. (Ideally the choice of exercise should be tailored to reflect the kinds of private events in the client's OCD that he is seeking to control.)

THE ALTERNATIVE TO CONTROL: WILLINGNESS

At this point the pitfalls of the control agenda have been made clear and therapy then progresses to consider the alternative—acceptance. However, it is preferable to use the term "willingness" because of the risk that the client may equate acceptance with tolerance, resignation, or grim forbearance. The following metaphor illustrates the point.

Two Scales

Present two scales, represented as two dials that can move up and down from 0 to 10, to the client. The first scale represents the severity of the obsession and its linked negative feelings (call it whatever experience best fits the client's presentation, such as guilt or disturbing thoughts; "anxiety" will be used in this illustration). The client desires the alleviation of anxiety and has asked for the therapist's help. But there is a second, hidden scale, called "willingness," which refers to how open the client is to fully experience his anxiety without trying to change, avoid, or escape it.

Explain that when trying to eliminate the anxiety, the client is unwilling to experience it; therefore, the willingness scale goes down to zero. However, if you're unwilling to have anxiety, then anxiety is something to get anxious about and fight against, and the anxiety dial can only be turned *up*—in the long term the OCD gets bigger and more entrenched. Even though the client's mind holds out that if willingness is at a minimum, anxiety will go away, her experience testifies to the opposite.

Therefore, treatment focuses on the willingness scale, which, given the client is seeking therapy, is likely set at low. Unlike the anxiety scale, it can be moved around because it reflects a *choice*. It needs to be set high. When this has been done, anxiety may be low...or high! But when willingness is low, anxiety *will* remain central and the client will continue to be embroiled in a vexing, unwinnable struggle.

The nature of willingness and commitment. Clarify to the client that willingness to experience distressing private events reflects making a choice: to try to control what you feel and lose control over your life, or to let go of trying to control this and so regain control over your life. It is not the same as ignoring or tolerating discomfort.

However, work needs to be done to prepare a platform from which the client can choose to experience painful thoughts and feelings. This is because, as you might tell your client, "There are four of us in the room—me and you and our two minds—and we can be certain that your mind will be fighting the move to increased willingness."

Introduce the client to the idea of advancing her willingness to experience distressing private events by making a behavioral commitment. This reflects choosing to pursue certain valued standards (one's *own*) in a set of defined situations, despite the OCD's sabotage attempts making life difficult. Also put forward the idea of being open to the feelings associated with occasionally slipping up in meeting the commitment, but choosing to persevere nonetheless.

Behavioral commitment exercises. These exercises, given as homework assignments, involve the client making a commitment to increasing her willingness to have the obsession (these are implemented from session 3 onward in Twohig's protocol). The aim is for defined, good-quality exercises in implementing value-directed behavior that is both important to the client and has been impeded by obsessions (the exercise can specify introducing an activity without doing a ritual for a daily defined time period, or limiting the daily frequency of rituals). Participation is not conditional on internal states. The aim is for the client to practice increasing her willingness to experience discomfiting private events, rather than anxiety habituation or eliminating the obsession. This is a routine homework assignment from this point on in the therapy, and the level of commitment enacted should gradually be increased.

Furthermore, the client may benefit from completing worksheets for homework such as the Daily Willingness Diary and the Clean and Dirty Discomfort Diary (included in Twohig, 2004). The former involves the client describing her experience after an obsession is triggered, with separate columns for describing the feelings, thoughts, and bodily sensations that constituted the experience and a final column for describing any coping strategy ("What did you do to handle…?").

The Clean and Dirty Discomfort Diary is a more sophisticated tool that draws on the distinction between "clean" and "dirty" discomfort (*clean*: "what life throws at you—what comes and goes as a result of living your life; what you can't get rid of by trying to control it" versus *dirty*: "upsetting thoughts and emotional discomfort that are caused by your attempt to control your feelings"). Instruct the client to complete the diary when she feels "stuck" with an obsession. The aim is to identify the ways in which the client "struggled" with the obsession (therefore, causing preventable suffering associated with dirty discomfort) and establish a basis for eliminating these.

SELF-AS-CONTEXT AND DEFUSION

The next sessions are typically aimed at changing the psychological function of the obsession from something threatening to just another mental event. To this end, introduce self-as-context and defusion exercises. The former represents a psychological posture best learned through practice. Defusion strategies are aimed at broadening the client's behavioral repertoire for interacting with the obsession, which reduces its believability, threat value, and impact. The following metaphors and exercises illustrate these concepts.

Passengers on the Bus

The aim of this exercise is to assist in defusion from challenging psychological content through objectification. Ask the client to imagine being a bus driver driving a bus full of passengers—this represents all of his experiences, thoughts, bodily states, memories, and so on. Some of the passengers are scary and thuggish, with knives and baseball bats. As the bus drives along, they threaten the driver, telling him if he doesn't drive where they want to go, they'll come up from the back, stand next to him, shout at him, and maybe even beat him up. The driver could (1) make deals with the passengers to keep them at the back—but then he has given them control of the direction the bus takes, and the passengers have the additional leverage of the deals that he made with them, or (2) stop the bus and try to throw the passengers off—but they're strong and would wrestle with him, he wouldn't have much success, and the bus would have stopped its journey and would be losing time, or (3) carry on the journey in the driver's chosen direction, while putting up with the passengers being unruly.

The metaphor aims to convey that when opting for 1 and 2 to get control over the passengers (thoughts and feelings), the driver has sacrificed control over the direction of the bus (the client's life direction), which is now controlled by them. In opting for 3, the driver has regained control of the direction—but the passengers will be noisier.

You can further elaborate on this by adding that the passengers, despite seeming scary and threatening to destroy the driver if he doesn't do as he is told, have never done anything worse than wave their arms, shout, and stamp their feet. Despite causing a ruckus, they can't actually make the driver do anything against his will. At later points in therapy, when the client is lapsing into experiential avoidance, the therapist can refer back to the metaphor by asking, "What is that passenger telling you now?"

Chessboard

Introduce the client to the metaphor of an infinite chessboard, where the pieces represent her thoughts and feelings. She is commanding the battle from the level of the pieces, aiming to beat the opponent. But there are problems here: she is trying to capture the opponent's pieces, but shortly after she puts them to the side, they suddenly appear back on the board. Areas of her mind have become hostile terrain; her experience is now her enemy. The pieces loom larger as the battle becomes more entrenched and central to her life. She is feeling hopeless and miserable as the conflict deepens and escalates.

At this point, turn the questions to the positioning of the self by asking, "If you're not the pieces, then where are you?" This leads the client to consider the idea of the self being represented by the board—which holds the pieces and defines their role and meaning. Even though in direct contact with the pieces, the board doesn't care about the game; it can watch the battle unfold, but the outcome doesn't matter. As the chessboard can exist without being harmed by the pieces, the obsessions can exist without harming the self.

From that point, the metaphor can become an anchoring point for reflection: "board level" designates a stance of *looking at* psychological content, rather than being enmeshed with it; the pieces are intrusions, feelings, and so on, not the self.

Language Reformatting

Present these exercises to the client as ways of altering language to reduce the pulling power of thinking ("how you are pulled into the battle zone/to the level of the chess pieces," see above), by drawing attention to what is really being said. These can be particularly helpful to introduce when the client is finding himself getting entangled in his OCD thinking.

In the "I Am Doing/Having" exercise, give the client the following language guidelines to use when communicating:

- Adopt "I" statements when describing having a private event.
- Describe the behavioral process clearly (remember that in ACT, language, or cognition, is considered to be behavior).
- Use verbs of doing or having, rather than being.

For example, have the client reformat "I am anxious" as "I am having a feeling of anxiety" (separating the self from the experience); "I am worried there could be germs on my hands" becomes "I am having thoughts that there could be germs on my hands and thoughts about what could go wrong and feelings of anxiety and guilt" (allowing distance to see reactions as just that).

The second exercise involves substituting "buts" with "ands," with the former being more at the chess piece level and the latter more at the board level. For example, the client would change "I want to go to work, but I am nervous" (*but* elevates the feeling into a reason for not choosing valued behavior) to "I want to go to work *and* I am feeling nervous" (both are true, but the latter is not getting in the way of the former).

Word Repetition

Ask the client to repeatedly say a word or phrase containing a feared meaning, such as "mistake" or "molester," which results in the experience of hearing the word morphing into meaningless noise (mistakemistakemistakemistakemistake...). This illustrates how literal meaning dominates in the ordinary use of language, but can easily be weakened through altering the context (repetition). In addition to the meaning behind the word, it is just a sound—it doesn't contain anything of substance.

Noting You Are the One Noticing

This experiential exercise, which takes about ten minutes, can be practiced in session and given for homework. The aim is to foster a sense of self-as-context that allows a context for defusion. Guide the client to mindfully reflect on her present experience (acknowledge body sensation, feelings, thoughts), finally noting that she is the one noticing (the "observer you")—the person she has been all her life. Then repeat these steps for different time points in the client's history: for "something that happened last year/when you were a teenager/when you were a child." Every time, the continuity of the observing self is affirmed. Finally, ask the client to contemplate the diverse multitude of body sensations, life roles, emotions, and thoughts she has experienced, and how the observing self transcends that. And have her notice how the *self* can adopt a posture of willingness—accepting of and connected to her experience, exercising her freedom to actively choose a value-guided direction in her actions.

Mindfulness

Mindfulness exercises, such as the Leaves on a Stream exercise (see Luoma, Hayes, & Walser, 2007), can be used to support defusion. Ask the client to close his eyes and picture being next to a stream. Guide him to imagine placing his thoughts gently on leaves floating down the stream. When hooked by a thought, have him notice what occurred, place the thought on a leaf, and allow it to go down the stream. If the client thinks, *I'm not doing it right*, that becomes the first thought to put on a leaf. The point is to learn to notice experiences, in the moment, without holding on to them.

At this point in treatment, hopefully the client has reduced her level of ritualizing, incentivized by the incremental benefits. Likely there has been substitution of compulsions with *valued action* (i.e., value-guided behavior). You can now increase the momentum for valued-guided behavior by doing structured work in this area in the final sessions.

VALUE-GUIDED BEHAVIOR

Clients can benefit from considering their goals and values (general life directions) in the following domains: romantic/family/friendship relationships, employment, education/personal development, recreation, spirituality, citizenship, and health behaviors. In each of these areas, prompt the client to consider what she would seek to attain and what defines an ideal situation: what kind of relationship, career, or training, and what hobbies or forms of participation, and why these matter to her.

Then have the client identify the qualities that embody desired behaviors (e.g., being patient) and specific behaviors that would define moving in a desired direction ("act in a patient way toward my partner when I am feeling stressed"). This material can be briefly summarized as a valued direction for each of the listed domains. Finally, the client can assess on a 0 to 10 scale how successfully she has lived according to the valued direction over the past two weeks or month and can highlight the domains that she considers important to work on. This material translates into increased behavioral commitments to live more consistently with important values.

Handy worksheets for assisting therapy work in these areas are included in Twohig (2004). Alternatively, the Valued Living Questionnaire (Wilson, Sandoz, & Kitchens, 2010; available in an appendix to their paper) offers a brief and practical strategy for rating the importance of different life domains and the degree of success with value-directed behavior in them. The following metaphor illustrates several relevant ACT concepts.

Joe the Neighbor

Have the client imagine that he is hosting a street party at his house, but then Joe the neighbor turns up. Joe's personal hygiene is poor, he has food stains on his clothes, and his teeth look like he infrequently cleans them. In addition, he tells crude jokes and goes off on boring tangents. Despite having said everyone is welcome, the host can (1) try to push Joe out of the house, but then he may fetch his dubious friends and try to get back in, in which case the host will have to guard the door and miss the party; (2) try to keep Joe in the kitchen so he doesn't mingle with the other guests, but then the host will have to spend all evening being bored by Joe; or (3) try to persuade him to clean up and put some deodorant on, but Joe won't have any of it and even if he did, he'd be a mess again in no time and that wouldn't stop the jokes!

In all of these responses, the host ends up missing or being distracted from the party—it's not life enhancing, it's tiring and ends up being all about Joe. The metaphor aims to convey that Joe represents unwanted thoughts and feelings, and that the client has a choice: He can try to eliminate these experiences (as in responses 1 through 3 above), but this is at a cost (such as when fighting one unwanted experience, others follow, such as Joe's friends). Or, he can adopt a posture of willingness to have Joe as his guest, which is distinct from his evaluation of him, and engage fully in the party—this option represents valued living!

RELAPSE PREVENTION

The last stage of ACT involves reviewing the therapy work and showing the client how to continue practicing key therapy insights of acceptance, defusion, and self-as-context, while maintaining and broadening her engagement in value-consistent action.

Final Remarks

In this book I have given an overview of the cutting-edge, contemporary evidence-based psychological treatments for OCD. Currently—due to the tireless efforts of researchers and clinicians—there is a menu of treatment options, both psychiatric and psychological, available for OCD, and it is very hard to believe that scarcely more than sixty years ago there was no effective treatment.

Five innovative CBT approaches have been described. There are important areas of overlap and convergence, but also clear differences. It is beyond the scope of this book to offer a critical comparison and review of all the approaches, but I have tried to give some pointers—where data allowed this—as to their strengths, applications, and limitations. One can envisage that future research will further clarify, refine, and distill specific treatment modules from the existing treatments, targeting particular content and OCD processes. Further consideration will allow optimal selection according to clearly defined client and treatment context characteristics and sequencing of modules.

At this point, ERP maintains its position as the first-line evidence-based psychological treatment. However, cognitive therapy approaches have successfully made inroads in presenting a coherent alternative or complementary account on which intervention may be predicated. This likely offers an additional benefit by allowing clients a choice of treatments, and by allowing focused targeting of problems with treatment engagement and adherence and of symptom domains where cognitive content and processes afford unique therapeutic leverage. A pressing question is whether third-wave approaches will be able to expand the range of existing treatments.

I hope that the range of models and approaches described increases your therapeutic tool set and expands your clinical repertoire by allowing you to more flexibly and creatively engage with this very complex condition. Doubtless, the following years will see important further advances. Much remains to be done. This reminds me of the dictum—Rome wasn't built in a day!

Acknowledgments

The writing of this book was a happy experience. I would like to thank the following people for making this possible: First, the staff at New Harbinger: Jess O'Brien, who kindled the ideas that eventually translated into this title, and Nicola Skidmore, Clancy Drake, and Vicraj Gill, who gave helpful editorial feedback. I described my copy editor, Rona Bernstein, to friends as "super-smart, super-organized, and super-friendly," and this is as good a description as any. Further, I would also like to thank Elizabeth Forrester for her invaluable feedback on chapters 1 through 4—Liz generously gave me hours of her time. Adrian Wells provided very helpful comments on chapter 6. Mike Twohig put me in contact with his student, Eric Lee, who gave valuable feedback about chapter 7.

I would like to acknowledge previous funding received from the NHS East of England to allow me to do a postdoctoral research fellowship, investigating CBT for OCD. During this time, Peter Jones kindly afforded me the opportunity to be a visiting researcher at the Department of Psychiatry, University of Cambridge. I would like to thank my former boss, Geraldine Owen, who enabled me to further develop my specialty in OCD. I extend my gratitude to Tim Dalgleish and colleagues at the Cambridge Clinical Research Centre for Affective Disorders, where I have been an associate for the duration of this book project.

Finally, I would like to thank my wife, Isobel, for her good humor, forbearance, and support during hours of writing and for her editorial feedback, and past and present clients with OCD who allowed me to learn from their experience.

APPENDIX 1

Worksheets

If desired, you can download copies of these worksheets at http://www.newharbinger.com/38952

Symptom Diary

Time	**Situation** (in which you experience the urge to ritualize)	**Anxiety/ discomfort in situation** (0–100)	**Describe the ritual**	**Duration** (minutes or number of rituals)

Exposure Recording Form

Client name: ______________________________ Date: ______________

Exposure practiced: ____

Minutes	0	5	10	15	20	25	30	35	40	45	50	55	60
Anxiety/ discomfort (0–100)													

Processing (record what the client has learned about the feared consequences of exposure):

Problems encountered:

Exposure Practice Form

Name: ______________________________ Date: __________________

Exposure to practice: ____

	Anxiety/discomfort (0–100)						
Date	Start	10 min.	20 min.	30 min.	40 min.	50 min.	60 min.

Lessons learned:

Problems encountered:

Behavioral Experiment Worksheet

Name: ______________________________ Date: ____________________

Describe the experiment:

What is the experiment aiming to test?

How much do you believe your negative predictions on a scale of 0 to 100?

How exactly would you know if your predictions came true?

What exactly happened? Did your predictions come true?

How did your anxiety/distress at the start of the experiment compare with it at the end?

What have you learned?

Intrusive Thoughts Worksheet (with Instructions)

Situation/ intrusion-trigger Describe what triggered the intrusive thought.	***Intrusive thought*** Describe the unwanted thought/ image/urge.	***OCD interpretation*** Write how you interpreted the thought(s)/the meaning you attached to it. How much do you believe this from 0–100?	***Emotion*** What feelings did you experience?	***OCD coping strategies*** Describe any coping by performing rituals, neutralization, or through avoidance.	***Helpful interpretation and response*** What is a helpful, realistic way of interpreting the thought? How much do you believe this from 0–100? How would this allow you to act?

Intrusive Thoughts Worksheet (with Instructions)

Situation/ intrusion-trigger	*Intrusive thought*	*OCD interpretation*	*Emotion*	*OCD coping strategies*	*Helpful interpretation and response*

APPENDIX 2

Intrusions Reported by an Unselected Student Sample

The column on the left shows the unwanted thought/image/impulse, and the columns on the right show the percent of women and men who reported it (from Purdon & Clark, 1993). You can download a copy of this questionnaire at http://www.newharbinger.com/38952.

Questionnaire Item	*Female %*	*Male %*
Driving into a window	13	16
Running car off the road	64	56
Hitting animals or people with car	46	54
Swerving into traffic	55	52
Smashing into objects	27	40
Slitting wrist/throat	20	22
Cutting off finger	19	16
Jumping off a high place	39	46
Fatally pushing a stranger	17	34
Fatally pushing a friend	9	22
Jumping in front of train/car	25	29
Pushing a stranger in front of train/car	8	20

Questionnaire Item	Female %	Male %
Pushing family in front of train/car	5	14
Hurting strangers	18	48
Insulting strangers	50	59
Bumping into people	37	43
Insulting authority figure	34	48
Insulting family	59	55
Hurting family	42	50
Choking family member	10	22
Stabbing family member	6	11
Accidentally leaving heat/stove on	79	66
Leaving home unlocked and intruder gets in	77	69
Leaving taps on leading to flooding	28	24
Swearing in public	30	34
Breaking wind in public	31	49
Throwing something	28	26
Causing a public scene	47	43
Scratching car paint	26	43
Breaking a window	26	43
Wrecking something	32	33
Shoplifting	27	33
Grabbing money	21	38
Holding up bank	6	32
Having sex with unacceptable person	48	63

Questionnaire Item	Female %	Male %
Having sex with authority figure	38	63
Fly/blouse undone	27	40
Kissing authority figure	37	44
Exposing myself	9	21
Acts going against sexual preference	19	20
Authority figures naked	42	54
Strangers naked	51	80
Having sex in public	49	78
A disgusting sex act	43	52
Catching sexually transmitted disease	60	43
Contamination from doors	35	24
Contamination from phones	28	18
Getting fatal disease from strangers	22	19
Transmitting a fatal disease	25	17
Giving everything away	52	43
Removing all the dust from the floor	35	24
Dirt in unseen places	41	29

Reference List

Aardema, F., & O'Connor, K. (2007). The menace within: Obsessions and the self. *Journal of Cognitive Psychotherapy: An International Quarterly, 21,* 182–197.

Aardema, F., O'Connor, K. P., Delorme, M., & Audet, J. (2017). The inference-based approach (IBA) to the treatment of obsessive-compulsive disorder: An open trial across symptom subtypes and treatment-resistant cases. *Clinical Psychology and Psychotherapy, 24,* 289–301.

Aardema, F., Wu, K. D., Careau, Y., O'Connor, K., Julien, D., & Dennie, S. (2010). The expanded version of the Inferential Confusion Questionnaire: Further development and validation in clinical and non-clinical samples. *Journal of Psychopathology and Behavioral Assessment, 32,* 448–462.

Abramovitch, A., Abramowitz, J., & Mittelman, A. (2013). The neuropsychology of adult obsessive–compulsive disorder: A meta-analysis. *Clinical Psychology Review, 33,* 1163–1171.

Abramowitz, J. (1996). Variants of exposure and response prevention in the treatment of obsessive-compulsive disorder. *Behavior Therapy, 27,* 583–600.

Abramowitz, J. (2006). *Obsessive-compulsive disorder.* Cambridge, MA: Hogrefe & Huber.

Abramowitz, J., & Deacon, B. J. (2006). Psychometric properties and construct validity of the Obsessive–Compulsive Inventory—Revised: Replication and extension with a clinical sample. *Anxiety Disorders, 20,* 1016–1035.

Abramowitz, J., Deacon, B., Olatunji, B., Wheaton, M., Berman, N., Losardo, D.,... Hale, L. R. (2010). Assessment of obsessive-compulsive symptom dimensions: Development and evaluation of the Dimensional Obsessive-Compulsive Scale. *Psychological Assessment, 22,* 180–198.

Abramowitz, J., & Jacoby, R. (2015). *Obsessive-compulsive disorder in adults.* Boston, MA: Hogrefe.

Abrantes, A., Strong, D. C., Cohn, A., Cameron, A., Greenberg, B., Mancebo, M., & Brown, R. A. (2009). Acute changes in obsessions and compulsions following moderate-intensity aerobic exercise among patients with obsessive-compulsive disorder. *Journal of Anxiety Disorders, 23,* 923–927.

American Psychiatric Association. (2013). *Diagnostic and statistical manual of mental disorders* (5th ed.). Washington, DC: Author.

Amir, N., Freshman, M., Ramsey, B., Neary, E., & Brigidi, B. (2001). Thought-action fusion in individuals with OCD symptoms. *Behaviour Research and Therapy, 39,* 765–776.

Arntz, A., Rauner, M., & van den Hout, M. (1995). "If I feel anxious, there must be danger": Ex-consequentia reasoning in inferring danger in anxiety disorders. *Behaviour Research and Therapy, 33,* 917–25.

Baer, L. (1991). *Getting control: Overcoming your obsessions and compulsions.* Boston, MA: Little, Brown, & Co.

Baker, A., Mystkowski, J., Culver, N., Yi, R., Mortazavi, A., & Craske, M. (2010). Does habituation matter? Emotional processing theory and exposure therapy for acrophobia. *Behaviour Research and Therapy, 48,* 1139–43.

Bandura, A. (1982). Self-efficacy mechanism in human agency. *American Psychologist, 37,* 122–147.

Barcaccia, B., Tenore, K., & Mancini, F. (2015). Early childhood experiences shaping vulnerability to obsessive-compulsive disorder. *Clinical Neuropsychiatry, 12,* 141–147.

Beck, A. (1976). *Cognitive therapy and the emotional disorders.* New York, NY: International Universities Press.

Beck, J. (1995). *Cognitive therapy: Basics and beyond.* New York, NY: Guilford Press.

Beck, J. (2011). *Cognitive behavior therapy: Basics and beyond* (2nd ed.). New York, NY: Guilford Press.

Berry, L., & Laskey, B. (2012). A review of obsessive intrusive thoughts in the general population. *Journal of Obsessive-Compulsive and Related Disorders, 1,* 125–132.

Bloch, M., McGuire, J., Landeros-Weisenberger, A., Leckman, J., & Pittenger, C. (2010). Meta-analysis of the dose-response relationship of SSRI in obsessive-compulsive disorder. *Molecular Psychiatry, 15,* 850–855.

Bluett, E. J., Homan, K. J., Morrison, K. L., Levin, M. E., & Twohig, M. P. (2014). Acceptance and commitment therapy for anxiety and OCD spectrum disorders: An empirical review. *Journal of Anxiety Disorders, 28,* 612–624.

Browne, H., Gair, S., Scharf, J., & Grice, D. (2014). Genetics of obsessive-compulsive disorder and related disorders. *Psychiatric Clinics of North America, 37*(3), 319–335.

Careau, Y., O'Connor, K., Turgeon, L., & Freeston, M. (2012). Childhood experiences and adult beliefs in obsessive-compulsive disorder: Evaluating a specific etiological model. *Journal of Cognitive Psychotherapy: An International Quarterly, 26,* 236–256.

Clark, D. (2004). *Cognitive-behavioral therapy for OCD.* New York, NY: Guilford Press.

Clark, D., & Beck, A. (2010). *Cognitive therapy of anxiety disorders: Science and practice.* New York, NY: Guilford Press.

Craske, M., Treanor, M., Conway, C., Zbozinek, T., & Vervliet, B. (2014). Maximizing exposure therapy: An inhibitory learning approach. *Behaviour Research and Therapy, 58*, 10–23.

Damasio, A. (2005). Descartes' error: Emotion, reason, and the human brain. New York, NY: Penguin.

Dek, E. C. P., van den Hout, M. A., Engelhard, I. M., Giele, C. L., & Cath, D. C. (2015). Perseveration causes automatization of checking behavior in obsessive-compulsive disorder. *Behaviour Research and Therapy, 71*, 1–9.

Dennett, D. C. (2014). The evolution of reasons. In B. Bashour & H. D. Muller (Eds.), *Contemporary philosophical naturalism and its implications* (pp. 47–62). New York, NY: Routledge.

De Silva, P. (2003). Obsessions, ruminations and covert compulsions. In R. Menzies & P. De Silva (Eds.), *Obsessive-compulsive disorder: Theory, research and treatment* (pp. 195–208). Chichester, UK: Wiley.

Dold, M., Aigner, M., Lanzenberger, R., & Kasper, S. (2015). Antipsychotic augmentation of serotonin reuptake inhibitors in treatment-resistant obsessive-compulsive disorder: An update meta-analysis of double-blind, randomized, placebo-controlled trials. *International Journal of Neuropsychopharmacology, 18*, 1–11.

Eisen, J., Phillips, K., Baer, L., Beer, D., Atala, K., & Rasmussen, S. (1998). The Brown Assessment of Beliefs Scale: Reliability and validity. *American Journal of Psychiatry, 155*, 102–108.

Eisen, J., Yip, A., Mancebo, M., Pinto, A., & Rasmussen, S. (2010). Phenomenology of obsessive-compulsive disorder. In D. Stein, E. Hollander, & B. Rothbaum (Eds.), *Textbook of anxiety disorders* (2nd ed., pp. 261–286). Arlington, VA: American Psychiatric Publishing.

Ellis, A., & Ellis, D. (2011). *Rational emotive behavior therapy.* Washington, DC: American Psychological Association.

Ellis, T., & Newman, C. (1996). *Choosing to live: How to defeat suicide through cognitive therapy.* Oakland, CA: New Harbinger.

Fineberg, N., & Craig, K. (2010). Pharmacotherapy for obsessive-compulsive disorder. In D. Stein, E. Hollander, & B. Rothbaum (Eds.), *Textbook of anxiety disorders* (2nd ed., pp. 311–337). Arlington, VA: American Psychiatric Publishing.

Fisher, P. (2009). Obsessive-compulsive disorder: A comparison of CBT and the metacognitive approach. *International Journal of Cognitive Therapy, 2*, 107–122.

Fisher, P., & Wells, A. (2009). *Metacognitive therapy.* Hove, UK: Routledge.

Flaxman, P., Blackledge, J., & Bond, F. (2011). *Acceptance and commitment therapy.* Hove, UK: Routledge.

Flessner, C. A., Conelea, C. A., Woods, D. W., Franklin, M. E., Keuthen, N. J., & Cashin, S. E. (2008). Styles of pulling in trichotillomania: Exploring differences in symptom severity, phenomenology, and functional impact. *Behaviour Research and Therapy, 46*, 345–357.

Foa, E., Huppert, J., Leiberg, S., Langer, R., Kichic, R., Hajcak, G., & Salkovskis, P. M. (2002). The Obsessive-Compulsive Inventory: Development and validation of a short version. *Psychological Assessment, 14*, 485–96.

Foa, E., & Kozak, M. (1986). Emotional processing of fear: Exposure to corrective information. *Psychological Bulletin, 99*, 20–35.

Foa, E., & Kozak, M. (1996). Psychological treatment for obsessive-compulsive disorder. In M. R. Mavissakalian & R. F. Prien (Eds.), *Long-term treatments of anxiety disorders* (pp. 285–310). Washington, DC: American Psychiatric Press.

Foa, E., Yadin, E., & Lichner, T. (2012). *Exposure and response (ritual) prevention for obsessive compulsive disorder*. New York, NY: OUP.

Fonseka, T., Richter, M., & Müller, D. (2014). Second generation antipsychotic-induced obsessive-compulsive symptoms in schizophrenia: A review of the experimental literature. *Current Psychiatry Reports, 16*, 510.

Frost, R., & Steketee, G. (2002). *Cognitive approaches to obsessions and compulsions: Theory, assessment and treatment*. Oxford, UK: Elsevier.

Gaudiano, B. (2011). Evaluating acceptance and commitment therapy: An analysis of a recent critique. *International Journal of Behavioral Consultation and Therapy, 7*, 54–65.

Goodman, W. K., Price, L. H., Rasmussen, S. A., Mazure, C., Fleischmann, R. L., Hill, C. L.,…Charney, D. S. (1989). The Yale-Brown Obsessive Compulsive Scale: I. Development, use and reliability. *Archives of General Psychiatry, 46*, 1006–1011.

Grant, J., Chamberlain, S., & Odlaug, B. (2014). *Clinical guide to obsessive compulsive and related disorders*. New York, NY: Oxford University Press.

Grant, J., Mancebo, M., Pinto, A., Williams, K., Eisen, J., & Rasmussen, S. (2007). Late-onset obsessive compulsive disorder: Clinical characteristics and psychiatric comorbidity. *Psychiatry Research, 152*, 21–27.

Harris, R. (2009). *ACT made simple*. Oakland, CA: New Harbinger.

Hayes, S. C., Batten, S., Gifford, E., Wilson, K. G., Afari, N., & McCurry, S. M. (1999). *Acceptance and commitment therapy: A manual for the treatment of emotional avoidance* (2nd ed.). Reno, NV: Context Press.

Hayes, S., Levin, M., Plumb-Vilardaga, J., Villatte, J., & Pistorello, J. (2013). Acceptance and commitment therapy and contextual behavioral science: Examining the progress of a distinctive model of behavioral and cognitive therapy. *Behavior Therapy, 44*, 180–198.

Hayes, S., Strosahl, K., & Wilson, K. G. (1999). *Acceptance and commitment therapy: An experiential approach to behavior change*. New York, NY: Guilford Press.

Hayes, S., Strosahl, K., & Wilson, K. (2012). *Acceptance and commitment therapy: The process and practice of mindful change* (2nd ed.). New York, NY: Guilford Press.

Hayes, S., Wilson, K., Gifford, E., Follette, V., & Strosahl, K. (1996). Experiential avoidance and behavioral disorders: A functional dimensional approach to diagnosis and treatment. *Journal of Consulting and Clinical Psychology, 64*, 1152–1168.

Hedman, E., Ljótsson, B., Andersson, G., Rück, C., & Andersson, E. (2017). Health anxiety in obsessive compulsive disorder and obsessive compulsive symptoms in severe health anxiety: An investigation of symptom profiles. *Journal of Anxiety Disorders, 45*, 80–86.

Heider, F., & Simmel, M. (1944). An experimental study of apparent behavior. *The American Journal of Psychology, 57*, 243–259.

Hirschtritt, M., Bloch, M., & Mathews, C. (2017). Obsessive-compulsive disorder: Advances in diagnosis and treatment. *JAMA, 317*, 1358–1367.

Issari, Y., Jakubovski, E., Bartley, C., Pittenger, C., & Bloch, M. (2016). Early onset of response with selective serotonin reuptake inhibitors in obsessive-compulsive disorder. *Journal of Clinical Psychiatry, 77*, 605–611.

Jenike, M., Baer, L., & Minichiello, W. (1998). *Obsessive-compulsive disorders: Practical management* (3rd ed.). St Louis, MN: Elsevier.

Julien, D., O'Connor, K. P., & Aardema, F. (2007). Intrusive thoughts, obsessions, and appraisals in obsessive-compulsive disorder: A critical review. *Clinical Psychology Review, 27*, 366–383.

Julien, D., O'Connor, K., & Aardema, F. (2016). The inference-based approach to obsessive-compulsive disorder: A comprehensive review of its etiological model, treatment efficacy, and model of change. *Journal of Affective Disorders, 202*, 187–96.

Lee, H., & Kwon, S. (2003). Two different types of obsession: autogenous obsessions and reactive obsessions. *Behaviour Research and Therapy, 41*, 11–29.

Linden, D. (2006). How psychotherapy changes the brain—the contribution of functional neuroimaging. *Molecular Psychiatry, 11*, 528–38.

Linehan, M. M. (1993). *Cognitive-behavioral treatment of borderline personality disorder.* New York, NY: Guilford Press.

Luoma, J., Hayes, S., & Walser, R. (2007). *Learning ACT: An acceptance and commitment therapy skills-training manual for therapists.* Oakland, CA: New Harbinger.

Maltby, N., & Tolin, D. (2005). A brief motivational intervention for treatment-refusing OCD patients. *Cognitive Behaviour Therapy, 34*, 176–184.

Mataix-Cols, D., Fernández de la Cruz, L., Monzani, B., Rosenfield, D., Andersson, E., Pérez-Vigil, A.,…Thuras, P. (2017). D-cycloserine augmentation of exposure-based cognitive behavior therapy for anxiety, obsessive-compulsive, and posttraumatic stress disorders. *JAMA Psychiatry, 74*, 501–510.

Mataix-Cols, D., Frost, R., Pertusa, A., Clark, L., Saxena, S., Leckman, J.,...Wilhelm, S. (2010). Hoarding disorder: a new diagnosis for DSM-V? *Depression and Anxiety, 27*, 556–572.

Menchón, J. (2012). Assessment. In J. Zohar (Ed.), *Obsessive compulsive disorder: Current science and clinical practice* (pp. 3–30). Chichester, UK: Wiley.

Menzies, L., Chamberlain, S. R., Laird, A. R., Thelen, S. M., Sahakian, B. J., & Bullmore, E. T. (2008). Integrating evidence from neuroimaging and neuropsychological studies of obsessive-compulsive disorder: The orbitofronto-striatal model revisited. *Neuroscience & Biobehavioral Reviews, 32*, 525–49.

Meyer, V. (1966). Modification of expectations in cases with obsessional rituals. *Behaviour Research and Therapy, 4*, 273–280.

Miller, W., & Rollnick, S. (2002). *Motivational interviewing* (2nd ed.). New York, NY: Guilford Press.

Moffitt, T., Caspi, A., Taylor, A., Kokaua, J., Milne, B., Polanczyk, G., & Poulton, R. (2010). How common are common mental disorders? Evidence that lifetime prevalence rates are doubled by prospective versus retrospective ascertainment. *Psychological Medicine, 40*, 899–909.

Morillo, C., Belloch, A., & García-Soriano, G. (2007). Clinical obsessions in obsessive-compulsive patients and obsession-relevant intrusive thoughts in non-clinical, depressed and anxious subjects: Where are the differences? *Behaviour Research and Therapy, 45*, 1319–1333.

Mowrer, O. (1960). *Learning theory and behavior.* Hoboken, NJ: Wiley.

Muris, P., Merckelbach, H., & Clavan, M. (1997). Abnormal and normal compulsions. *Behaviour Research and Therapy, 35*, 249–52.

National Institute for Health and Care Excellence. (2005, November). *Obsessive-compulsive disorder and body dysmorphic disorder: treatment—Clinical guideline [CG31].* Retrieved from https://www.nice.org.uk

Normann, N., van Emmerik, A. A. P., & Morina, N. (2014). The efficacy of metacognitive therapy for anxiety and depression: A meta-analytic review. *Depression and Anxiety, 31*, 402–411.

Obsessive Compulsive Cognitions Working Group. (2005). Psychometric validation of the obsessive belief questionnaire and interpretation of intrusions inventory—Part 2: Factor analyses and testing of a brief version. *Behaviour Research and Therapy, 43*, 1527–42.

O'Connor, K. (2005). *Cognitive-behavioral management of tic disorders.* Chichester, UK: Wiley.

O'Connor, K., & Aardema, F. (2012). *The clinician's handbook for obsessive compulsive disorder: Inference-based therapy.* Chichester, UK: Wiley.

O'Connor, K. P., Aardema, F., & Pélissier, M. (2005). *Beyond reasonable doubt: Reasoning processes in obsessive-compulsive disorder and related disorders*. Chichester, UK: Wiley.

O'Connor, K., Koszegi, N., Aardema, F., van Niekerk, J., & Taillon, A. (2009). An inference-based approach to treating obsessive-compulsive disorder. *Cognitive and Behavioral Practice, 16*, 420–429.

Olatunji, B. M, Berg, H., Cox, R. C., & Billingsley, A. (2017). The effects of cognitive reappraisal on conditioned disgust in contamination-based OCD: An analogue study. *Journal of Anxiety Disorders, 51*, 86–93.

Ong, C. W., Clyde, J., Bluett, E., Levin, M., & Twohig, M. (2016). Dropout rates in exposure with response prevention for obsessive-compulsive disorder: What do the data really say? *Journal of Anxiety Disorders, 40*, 8–17.

Pace, S. M., Thwaites, R, & Freeston, M. H. (2011). Exploring the role of external criticism in obsessive compulsive disorder: A narrative review. *Clinical Psychology Review, 31*, 361–370.

Padder, T. A. (2015). *Practical guide to psychiatric medications: Simple, concise, & up-to-date*. Clarksville, MD: MTP Publishing.

Padesky, C. (1994). Schema change processes in cognitive therapy. *Clinical Psychology and Psychotherapy, 1*, 267–278.

Padesky, C., & D. Greenberger. (1995). *Mind over mood: Change how you feel by changing the way you think*. New York, NY: Guilford Press.

Pauls, D. L., Abramovitch, A., Rauch, S., & Geller, D. (2014). Obsessive–compulsive disorder: An integrative genetic and neurobiological perspective. *Nature Reviews Neuroscience, 15*, 410–424.

Pinto, A., Mancebo, M., Eisen, J., Pagano, M., & Rasmussen, S. (2006). The Brown Longitudinal Obsessive Compulsive Study: Clinical features and symptoms of the sample at intake. *Journal of Clinical Psychiatry, 67*, 703–711.

Pirie, M. (2006). *How to win every argument: The use and abuse of logic*. London, UK: Continuum.

Plumb, J., Stewart, I., Dahl, J., & Lundgren, T. (2009). In search of meaning: Values in modern clinical behavior analysis. *The Behavior Analyst, 32*, 85–103.

Pollard, C. (2007). Treatment readiness, ambivalence and resistance. In M. Antony, C. Purdon, & L. Summerfeldt (Eds.), *Psychological treatment of obsessive-compulsive disorder* (pp. 61–78). Washington, DC: American Psychological Association.

Polman, A., O'Connor, K., & Huisman, M. (2011). Dysfunctional belief-based subgroups and inferential confusion in obsessive–compulsive disorder. *Personality and Individual Differences, 50*, 153–158.

Purdon, C. (2007). Cognitive therapy for obsessive-compulsive disorder. In M. Antony, C. Purdon, & L. Summerfeldt (Eds.), *Psychological treatment of obsessive-compulsive disorder* (pp. 111–146). Washington, DC: American Psychological Association.

Purdon, C., & Clark, D. (1993). Obsessive intrusive thoughts in nonclinical subjects. Part 1. Content and relation with depressive, anxious and obsessional symptoms. *Behaviour Research and Therapy, 31*, 713–720.

Purdon, C., & Clark, D. (2002). The need to control thoughts. In R. Frost & G. Steketee (Eds.), *Cognitive approaches to obsessions and compulsions: Theory, assessment and treatment* (pp. 29–43). Oxford, UK: Elsevier.

Purdon, C., & Clark, D. (2005). *Overcoming obsessive thoughts.* Oakland, CA: New Harbinger.

Rachman, S. (2003a). Primary obsessional slowness. In R. Menzies & P. De Silva (Eds.), *Obsessive-compulsive disorder: Theory, research and treatment* (pp. 181–194). Chichester, UK: Wiley.

Rachman, S. (2003b). *The treatment of obsessions.* Oxford, UK: Oxford University Press.

Rachman, S. (2004). Fear of contamination. *Behaviour Research and Therapy, 42*, 1227–55.

Rachman, S., & De Silva, P. (1978). Abnormal and normal obsessions. *Behaviour Research and Therapy, 16*, 233–248.

Rachman, S., & Hodgson, R. (1980). *Obsessions and compulsions.* Hemel Hempstead, UK: Prentice Hall.

Radomsky, A., Gilchrist, P., & Dussault, D. (2006). Repeated checking really does cause memory distrust. *Behaviour Research and Therapy, 44*, 305–16.

Reuman, L., Jacoby, R., Blakey, S., Riemann, B., Leonard, R., & Abramowitz, J. (2017). Predictors of illness anxiety symptoms in patients with obsessive compulsive disorder. *Psychiatry Research, 256*, 417–422.

Rosa-Alcázar, A., Sánchez-Meca, J., Gómez-Conesa, A., & Marín-Martínez, F. (2008). Psychological treatment of obsessive–compulsive disorder: A meta-analysis. *Clinical Psychology Review, 28*, 1310–1325.

Rosario-Campos, M. C., Miguel, E.C., Quatrano, S., Chacon, P., Ferrao, Y., Findley, D.,...Leckman, J. F. (2006). The Dimensional Yale-Brown Obsessive-Compulsive Scale (DY-BOCS): an instrument for assessing obsessive-compulsive symptom dimensions. *Molecular Psychiatry, 11*, 495–504.

Rowa, K., Gifford, S., McCabe, R., Milosevic, I., Antony, M., & Purdon, C. (2014). Treatment fears in anxiety disorders: Development and validation of the Treatment Ambivalence Questionnaire. *Journal of Clinical Psychology, 70*, 979–993.

Ruscio, A., Stein, D., Chiu, W., & Kessler, R. (2010). The epidemiology of obsessive-compulsive disorder in the National Comorbidity Survey Replication. *Molecular Psychiatry, 15*, 53–63.

Salkovskis, P. (1999). Understanding and treating obsessive-compulsive disorder. *Behaviour Research and Therapy, 37*, S29–S52.

Salkovskis, P., & Forrester, E. (2002). Responsibility. In R. Frost & G. Steketee (Eds.), *Cognitive approaches to obsessions and compulsions: Theory, assessment and treatment* (pp. 45–61). Oxford, UK: Elsevier.

Salkovskis, P., & Freeston, M. (2001). Obsessions, compulsions, motivation, and responsibility for harm. *Australian Journal of Psychology, 53*, 1–6.

Salkovskis, P., Shafran, R., Rachman, S., & Freeston, M. H. (1999). Multiple pathways to inflated responsibility beliefs in obsessional problems: Possible origins and implications for therapy and research. *Behaviour Research and Therapy, 37*, 1055–1072.

Schruers, K., Koning, K., Luermans, J., Haack, M., & Griez, E. (2005). Obsessive–compulsive disorder: A critical review of therapeutic perspectives. *Acta Psychiatrica Scandinavica*, 261–271.

Segal, Z. V., Williams, J. M. G., & Teasdale, J. D. (2013). *Mindfulness-based cognitive therapy for depression* (2nd ed.). New York, NY: Guilford Press.

Simpson, H. B., Zuckoff, A., Page, J., Franklin, M., & Foa, E. (2008). Adding motivational interviewing to exposure and ritual prevention for obsessive-compulsive disorder: An open pilot trial. *Cognitive Behaviour Therapy, 37*, 38–49.

Steketee, G. (1999). *Overcoming obsessive-compulsive disorder: A behavioral and cognitive protocol for the treatment of OCD.* Oakland, CA: New Harbinger.

Steketee, G., & Van Noppen, B. (2003). Family approaches to treatment for obsessive compulsive disorder. *Revista Brasileira de Psiquiatria, 25*, 43–50.

Stetka, B. (2016). In search of the optimal brain diet. *Scientific American Mind, 27*, 26–33.

Summerfeldt, L. J. (2004). Understanding and treating incompleteness in obsessive-compulsive disorder. *Journal of Clinical Psychology, 60*, 1155–1168.

Sumner, J., Noack, C., Filoteo, J., Maddox, W., & Saxena, S. (2016). Neurocognitive performance in unmedicated patients with hoarding disorder. *Neuropsychology, 30*, 157–168.

Thordarson, D. S., Radomsky, A., Rachman, S., Shafran, R., Sawchuk, C., & Hakstian, A. (2004). The Vancouver Obsessional Compulsive Inventory (VOCI). *Behaviour Research and Therapy, 42*, 1289–314.

Tolin, D., Abramowitz, J., & Diefenbach, G. (2005). Defining response in clinical trials for obsessive-compulsive disorder: A signal detection analysis of the Yale-Brown obsessive compulsive scale. *Journal of Clinical Psychiatry, 66*, 1549–57.

Twohig, M. (2004). ACT for OCD: Abbreviated treatment manual. Retrieved from https://contextualscience.org

Twohig, M. (2009). The application of acceptance and commitment therapy to obsessive-compulsive disorder. *Cognitive and Behavioral Practice, 16*, 18–28.

Twohig, M. (2012). Introduction: The basics of acceptance and commitment therapy. *Cognitive and Behavioral Practice, 19*, 499–507.

Twohig, M. P., Hayes, S. C., Plumb, J. C., Pruitt, L. D., Collins, A. B., Hazlett-Stevens, H., & Woidneck, M. R. (2010). A randomized clinical trial of acceptance and commitment therapy vs. progressive relaxation training for obsessive compulsive disorder. *Journal of Consulting and Clinical Psychology, 78*, 705–716.

van den Hout, M. A., Engelhard, I. M., Smeets, M., Dek, E. C. P., Turksma, K., & Saric, R. (2009). Uncertainty about perception and dissociation after compulsive-like staring: Time course of effects. *Behaviour Research and Therapy, 47*, 535–539.

van den Hout, M., & Kindt, M. (2003). Repeated checking causes memory distrust. *Behaviour Research and Therapy, 41*, 301–16.

van der Heiden, C., Van Rossen, K., Dekker, A., Damstra, M., & Deen, M. (2016). Metacognitive therapy for obsessive-compulsive disorder: A pilot study. *Journal of Obsessive-Compulsive and Related Disorders, 9*, 24–29.

van Niekerk, J. (2009). *Coping with obsessive-compulsive disorder.* Oxford, UK: Oneworld.

van Niekerk, J., Brown, G., Aardema, F., & O'Connor, K. (2014). Integration of inference-based therapy and cognitive-behavioral therapy for obsessive-compulsive disorder—A case series. *International Journal of Cognitive Therapy, 7*, 67–82.

van Oppen, P., & Arntz, A. (1994). Cognitive therapy for obsessive-compulsive disorder. *Behaviour Research and Therapy, 32*, 79–87.

Veale, D., Krebs, G., Heyman, I., & Salkovskis, P. (2009). Risk assessment and management in obsessive–compulsive disorder. *Advances in Psychiatric Treatment, 15*, 332–343.

Warwick, H. M., & Salkovskis, P. M. (1990). Unwanted erections in obsessive-compulsive disorder. *British Journal of Psychiatry, 157*, 919–21.

Wells, A. (1997). *Cognitive therapy of anxiety disorders*. Chichester, UK: Wiley.

Wells, A. (2005a). The metacognitive model of GAD: Assessment of meta-worry and relationship with DSM-IV generalized anxiety disorder. *Cognitive Therapy and Research, 29*, 107–121.

Wells, A. (2005b). Detached mindfulness in cognitive therapy: A metacognitive analysis and ten techniques. *Journal of Rational-Emotive & Cognitive-Behavior Therapy, 23*, 337–355.

Wells, A. (2009). *Metacognitive therapy for anxiety and depression.* New York, NY: Guilford.

Wells, A., & Cartwright-Hatton, S. (2004). A short form of the Metacognitions Questionnaire: Properties of the MCQ-30. *Behaviour Research and Therapy, 42*, 385–396.

Wells, A., Gwilliam, P., & Cartwright-Hatton, S. (2001). *The Thought Fusion Instrument.* Unpublished self-report scale, University of Manchester.

Wells, A., & Matthews, G. (1996). Modelling cognition in emotional disorder: The S-REF model. *Behaviour Research and Therapy, 32*, 867–870.

Wilhelm, S., & Steketee, G. (2006). *Cognitive therapy for obsessive-compulsive disorder.* Oakland, CA: New Harbinger.

Wilson, K., Sandoz, E., & Kitchens, J. (2010). The Valued Living Questionnaire: Defining and measuring valued action within a behavioral framework. *The Psychological Record, 60,* 249–272.

World Health Organization. (2001). *The world health report 2001—mental health: New understanding, new hope.* Geneva, Switzerland: World Health Organization.

Young, J., Klosko, J., & Weishaar, M. (2003). *Schema therapy: A practitioner's guide.* New York, NY: Guilford Press.

Zohar, J., Greenberg, B., & Denys, D. (2012). Obsessive-compulsive disorder. In T. E. Schlaepfer & C. B. Nemeroff (Eds.), *Handbook of clinical neurology: Neuropsychiatric disorders* (pp. 375–390). Amsterdam, Netherlands: Elsevier.

Jan van Niekerk, PhD, is a registered clinical psychologist practicing in Cambridge, United Kingdom. He is an associate at the Cambridge Clinical Research Centre for Affective Disorders and author of *Coping with Obsessive-Compulsive Disorder.*

Foreword writer **Christine Purdon, PhD, CPsych,** is associate professor of psychology at the University of Waterloo in Waterloo, ON, Canada; and consulting psychologist with the Anxiety Treatment and Research Clinic at St. Joseph's Healthcare in Hamilton, ON, Canada.

Index

A

B

C

D

E

F

G

H

I

J

K

L

M

N

O

P

Q

R

S

T

U

V

W

Y

Register your **new harbinger** titles for additional benefits!

When you register your **new harbinger** title—purchased in any format, from any source—you get access to benefits like the following:

- Downloadable accessories like printable worksheets and extra content
- Instructional videos and audio files
- Information about updates, corrections, and new editions

Not every title has accessories, but we're adding new material all the time.

Access free accessories in 3 easy steps:

1. Sign in at NewHarbinger.com (or **register** to create an account).
2. Click on **register a book**. Search for your title and click the **register** button when it appears.
3. Click on the **book cover or title** to go to its details page. Click on **accessories** to view and access files.

That's all there is to it!

If you need help, visit:

NewHarbinger.com/accessories